Guide to Technical Editing

GUIDE TO TECHNICAL EDITING

Discussion, Dictionary, and Exercises

ANNE EISENBERG

New York   Oxford
OXFORD UNIVERSITY PRESS
1992

Oxford University Press

Oxford   New York   Toronto
Delhi   Bombay   Calcutta   Madras   Karachi
Kuala Lumpur   Singapore   Hong Kong   Tokyo
Nairobi   Dar es Salaam   Cape Town
Melbourne   Auckland

and associated companies in
Berlin   Ibadan

Published by Oxford University Press, Inc.
200 Madison Avenue, New York, New York 10016

Oxford is a registered trademark of Oxford University Press

Library of Congress Cataloging-in-Publication Data
Eisenberg, Anne, 1942–
Guide to technical editing : discussion, dictionary, and exercises
/ Anne Eisenberg.
p. cm. ISBN 0-19-506306-6
1. Technical editing. I. Title.
T11.4.E38 1992 808'.0666—dc20 91-16696

9 8 7 6 5 4 3 2
Printed in the United States of America
on acid-free paper

PREFACE

This book is a beginner's guide to technical editing. It is written for people who want to learn how to edit either their own writing or the writing of others.

It has three parts:

1. A discussion of strategies for developing "an editor's eye."
2. A dictionary of basic terms that come into play in technical editing.
3. Exercises in editing.

STRATEGIES FOR EDITING

Some people can look at a document and immediately spot the problems—the subject-verb agreement errors, the inconsistencies in tense, the undefined terms. To many other people, though, such errors are invisible, for many reasons: (1) They may be trying to edit their own work, work with which they are so familiar that they are unable to assume the outside eye of the critic. (2) They may be dealing with a document with many interrelated problems, and end up overwhelmed by details that are difficult to separate and deal with individually. (3) They are not sure which problems represent correct and incorrect choices, and which represent variations in style that need only be made internally consistent.

What strategies are useful for beginners who want to develop an eye for editing? One effective way—a way used in this book—is to practice by sorting problems into categories, and then looking selectively within the category. Then the beginning editor either spots and corrects the problem, or consults the dictionary on, say, the use of "respectively" or the advisability of a comma, before returning to the task.

Editing skill builds quickly as beginners combine the reference information in a dictionary with practice isolating and correcting standard problems.

This book introduces typical problems that occur in technical documents under seven categories:

1. *Organization.* Is the structure of the document apparent? Do readers need a prefatory summary calling out the main points? Stronger lead sentences?
2. *Titles, headings, and subheadings.* Are the titles informative? Do headings mark main points and their order? What line position will be used for first-level subheadings? What size bullet for displayed lists?
3. *Definitions.* Are necessary terms defined? Do these definitions need to be clarified?
4. *Style.* Is a hyphen needed in the term *polynomial based descrambler?* A plural ending for *30 mL?* An apostrophe in *ICBMs?* Is the style concise and consistent?
5. *Tables.* Do columns need alignment? Are sources given? Do the tables display the data to their best effect?
6. *Figures.* Are captions and callouts consistent with the text? Are symbols that are used in the figures explained in the captions?
7. *Accuracy.* Do reference numbers match references? Are people's names and titles spelled correctly? Does *octyl dodecanol* appear as one word on page 3 but as two words on page 11?

Each strategy in the discussion is illustrated with two examples: a **before**—the writer's problem—and an **after**—the editor's solution. The explanation of the problem runs in a parallel column, as does the explanation of the solution.

The discussion of strategies is closely related to the Dictionary. Each technical term in the discussion is marked on first mention with an asterisk (*), meaning, "See Dictionary for definitions and explanations."

THE DICTIONARY

The Dictionary contains entries in three areas:

- Basic terms in grammar, usage, and style that come into play in technical editing, such as *eponyms, SI units, first person, abbreviations and acronyms, hyphen, active voice, qualifiers,* and *unnecessary words.*
- Printing and production terms like *format, font, pitch, point, flush left,* and *hanging paragraphs;* copy editing and proofreading marks and terms like *stet* and *indent one en.*
- Major types of technical writing, such as *instructions, proposals, abstracts, reports and scientific papers, results and discussions,* and *procedures.*

THE EXERCISES

Exercises in the earlier chapters have only one type of error. Later exercises contain multiple errors. The last exercise has a combination of problems in style, grammar, usage, organization, and accuracy in the text and in the accompanying tables, figures, and glossary.

This book is intended for students in technical writing classes interested in editing; for technical professionals who need to learn the rudiments of editing either for their own work or for the work of others who report to them; and for people from countries where conventions for technical editing differ from those in the United States.

All three parts of the book—the Dictionary, the discussion of strategies, and exercises—grew out of classes on behalf of the American Chemical Society or at Polytechnic University, or in response to readers of the *Technically Speaking* column in *IEEE Spectrum.*

The author would like to thank Philip Leininger of Oxford University Press, who suggested the combination of dictionary and workbook; and David Kiefer, *Chemical and Engineering News;* Margaret Eastman, Trudy E. Bell, and Ronald K. Jurgen, *IEEE Spectrum;* Jacqueline Kroschwitz, John Wiley & Sons; and Elizabeth Corcoran, *Scientific American.*

Brooklyn, N.Y.   A. E.
July 1991

CONTENTS

DISCUSSION

1/////

EDIT FOR ORGANIZATION THAT SHOWS MAIN POINTS AND SIGNIFICANCE

If writers of technical documents show a common weakness, it is in the density with which many recount their work.

This density poses problems for readers. Readers need to know main points *before* details; they need to understand the structure or path of the argument before the elaboration of supporting arguments. But many writers, caught in the technical loop of their expertise, begin the discussion at far too detailed a level for any but the most expert, leaving other readers to figure out the path of the argument, unable to distinguish main points from supporting ones without considerable effort.

How do editors intercede to make such arguments more accessible to readers? One tactic is to add or transpose text that shows main points, their significance, and their sequence before the details that support these points. This chapter discusses typical ways to bring out the underlying structure and emphases in a document.

STATE THE IDEA BEFORE ITS EXPLANATION.

Make the path of the argument clear by summing up main points before introducing detail. This will give the reader a framework for the supporting details of the argument.

You may be able to pair a statement of the main point with subheads giving the division.

not

Laboratory analyses indicate that natural carcinogens are found in a number of foods, including potatoes, parsley, and celery. Processing may add potential carcinogens to other foods. For example, the nitrates that are used to preserve meats such as bacon can combine with amines in the body to form

Problem: Author gives detail without first providing a framework for the detail.

nitrosamines, which are known carcinogens. Charcoal broiling of meat causes the formation of carcinogens called polycyclic hydrocarbons.

Population studies comparing the incidence of specific cancers in populations with different dietary patterns also have yielded valuable information. For example, the Japanese diet contains much less fat than the American diet, and the Japanese have a much lower incidence of breast cancer and colon cancer than Americans. This kind of evidence is suggestive but not conclusive. It is possible that some other element of the diet that is closely associated with fat is the causative agent, or that the overall diet, rather than individual nutrients, is responsible for the increased cancer risk.

but

Many foods contain substances that may influence cancer formation in humans. Evidence that diet is one factor in cancer comes from two entirely different kinds of data: laboratory analyses and population studies.

Laboratory analyses indicate that natural carcinogens are found in a number of foods, including potatoes, parsley, and celery. Processing may add potential carcinogens to other foods. For example, the nitrates that are used to preserve meats such as bacon can combine with amines in the body to form nitrosamines, which are known carcinogens. Charcoal broiling of meat causes the formation of carcinogens called polycyclic hydrocarbons.

Population studies comparing the incidence of specific cancers in populations with different dietary patterns have also yielded valuable data. For example, the Japanese diet contains much less fat than the American diet, and the Japanese have a much lower incidence of breast cancer and colon cancer than Americans. This kind of evidence is suggestive but not conclusive. It is possible that some other element of the diet that is closely associated with fat is the causative agent, or that the overall diet, rather than individual nutrients, is responsible for the increased cancer risk.

Solution: Editor adds a lead or opening paragraph summarizing the main points of the passage that follow: "Many foods contain substances that may influence cancer formation in humans. Evidence that diet is one factor in cancer comes from two entirely different kinds of data: laboratory analyses and population studies."

not

To: Date:
From: Distribution:
Re: New process to preserve books

This is an interesting new method. Using it, books are placed in an airtight container, which is then flushed with nitrogen to remove oxygen because oxygen inhibits polymerization. Then a monomer mixture consisting of ethyl acrylate and methyl methacrylate is introduced. An alkaline monomer is added to the mixture to deacidify the paper. After the monomer is given hours to distribute evenly within and between the books, the books are irradiated with a low dose of gamma rays from a cobalt-60 source. Gamma rays, which change the monomers into polymers, are used because they can

Problem: Writer plunges into detail of the new method without first summarizing the main point and its significance.

penetrate whole volumes and leave no residual radiation after treatment. The resulting polymer coats the fibers and strengthens the pages by forming bridges between the fibers, but the books are not fattened noticeably, and the paper still feels the same.

The traditional way to preserve books has been to deacidify them by spraying or soaking them with an alkali, at a cost of up to $340, rather than the new method, in which gamma rays irradiate the chemically treated books, coating the wood fibers with a polymer that strengthens them.

Books printed on paper made from wood pulp become brittle from acidification and are therefore in danger of disintegration. The prototype method, sponsored by the library and undertaken at the chemistry department of the University of Surrey, Guildford, for the past nine years, lowers preservation costs to about $8 per book.

Books are irradiated at the UK Atomic Energy Research Establishment, Harwell, Essex.

The library hopes to attract a company to set up a pilot plant that would test the research work more fully.

but

To: Date:
From: Distribution:
Re: Cheaper Way to Preserve Books

The British Library has developed a new, low-cost method to preserve books. The method, which uses gamma rays to irradiate chemically treated paper, costs only $8 per volume—a significant savings over the $340 price tag for the present method.

The standard way to preserve books made from wood pulp is to deacidify them by spraying or soaking with an alkali. This prevents the paper from becoming brittle from acidification and disintegration. The new technique, still in prototype, coats the wood fibers with a polymer that strengthens them.

The prototype was developed at the chemistry department at the University of Surrey, Guildford, which has been working on the project for the British Library over the past nine years.

Solution: Editor provides statement of the main point (new method that uses gamma rays to irradiate chemically treated paper) and its significance (costs $8/volume) before giving details of preservation process.

Details of the Process

The method is based on the creation of a polymer that coats the fibers and strengthens the pages by forming bridges between the fibers. The books are not thickened noticeably, and the paper feels the same. The process has three steps:

1. *Removing Oxygen.* Books are placed in an airtight container, which is then flushed with nitrogen to remove oxygen. This is a key step, as oxygen inhibits polymerization.
2. *Deacidifying the Paper.* A monomer mixture of ethyl acrylate and methyl methacrylate is introduced, and an alkaline monomer added to deacidify the paper before irradiation.

Editor summarizes salient features in process before giving details, then emphasizes each step in the method with subheads (*Removing Oxygen, Deacidifying Paper, Irradiating the Books*).

3. *Irradiating the Books.* After the monomer is given hours to distribute evenly within and between the books, the books are irradiated with a low dose of gamma rays from a cobalt-60 source at the UK Atomic Energy Research Establishment, Harwell, Essex. Gamma rays, which change the monomers into polymers, are used because they can penetrate whole volumes and leave no residual radiation after treatment.

The library hopes to attract a company to set up a pilot plant that will test the research work more fully.

not

This month's issue of the *Journal of Personality and Social Psychology* reports an unusual study. In the study, 26 pairs of male and female college students volunteered to watch scenes from horror movies and then rate how much they enjoyed the film and how desirable they found their companion.

The more distressed female companions seemed to be, the more attractive male subjects found them, and the more the men said they enjoyed the film. For the women, seeing a horror movie with men who showed no signs of fear led the women to rate the men as attractive, and to say that they enjoyed the film more.

The less appealing the man was initially, the more his attractiveness was raised in the eyes of his female companion by his showing no fear during the horror film.

Problem: Author presents fairly detailed discussion without first stating main point.

but

Adolescent men who can watch horror movies without looking upset are seen by their adolescent dates as more attractive because of their bravado, a new study reports.

In the study, which appears in this month's issue of the *Journal of Personality and Social Psychology,* 26 pairs of male and female college students volunteered to watch scenes from horror movies and then rate how much they enjoyed the film and how desirable they found their companion.

The more distressed female companions seemed to be, the more attractive male subjects found them, and the more the men said they enjoyed the film. For the women, seeing a horror movie with men who showed no signs of fear led the women to rate the men as attractive, and to say that they enjoyed the film more.

The less appealing the man was initially, the more his attractiveness was raised in the eyes of his female companion by his showing no fear during the horror film.

Solution: Editor adds opening sentence summing up passage: "Adolescent men who can watch horror movies without looking upset are seen by their adolescent dates as more attractive because of their bravado, a new study reports."

not

Cocaine abuse is widespread in the general population and has also increased among heroin-dependent persons, including those in methadone maintenance treatment programs. The many adverse medical consequences of cocaine abuse are aug-

Problem: No summary or abstract* that presents main points and their significance first.

*Terms marked with an asterisk are defined in the Dictionary.

mented by the combined use of cocaine and heroin. For example, dual addiction to intravenous cocaine and heroin may increase the risk of acquired immunodeficiency syndrome (AIDS), both through needle sharing and through the combined immunosuppressive effects of both drugs. Intravenous drug abuse was estimated to account for more than 30% of AIDS victims in the United States in 1988.

At present, there is no uniformly effective pharmacotherapy for cocaine abuse, and the dual abuse of cocaine plus heroin is an even more difficult treatment challenge. . . .

An ideal pharmacotherapy would be one that antagonized the reinforcing effects of cocaine and that had minimal adverse side effects or potential for abuse liability. The opioid mixed agonist-antagonist buprenorphine meets these criteria for the treatment of opiate abuse. Buprenorphine effectively suppressed heroin self-administration by heroin-dependent men during in-patient studies and blocked opiate effects for more than 24 hours. Cessation of buprenorphine treatment does not produce severe and protracted withdrawal signs and symptoms in man. Buprenorphine is safer than methadone because its antagonist component appears to prevent lethal overdose, even at approximately ten times the analgesic therapeutic dose. Buprenorphine is also effective for the out-patient detoxification of heroin-dependent persons.

Here we describe the effect of buprenorphine treatment on cocaine self-administration by rhesus monkeys. Cocaine effectively maintains operant responding, leading to its intravenous administration in primates, and it is well established that primates self-administer most drugs abused by man. The primate model of drug self-administration is a useful method for the prediction of drug abuse liability and can be used to evaluated new pharmacotherapies for drug abuse disorders.

Two male and three female adult rhesus monkeys (*Macaca mulatta*) with a 262 ± 79 day history of cocaine self-administration were studied. Each monkey was implanted with a double-lumen silicone rubber intravenous catheter under aseptic conditions to permit administration of buprenorphine or saline during cocaine self-administration. The intravenous catheter was protected by a custom-designed tether system (Spaulding Medical Products) that permits monkeys to move freely. Monkeys worked for food (1-g banana pellets) and for intravenous cocaine (0.05 or 0.10 mg per kilogram of body weight per injection) on the same operant schedule of reinforcement. . . . The total number of cocaine injections was limited to 80 per day to minimize the possibility of adverse drug effects. The nutritionally fortified diet of banana pellets was supplemented with fresh fruit, vegetables, biscuits, and multiple vitamins each day. . . .

We measured cocaine and food self-administration during 15 days of saline treatment and six successive 5-day periods of buprenorphine treatment. . . All animals reduced their cocaine self-administration significantly during buprenorphine treatment. On the first day of buprenorphine treatment, cocaine

Reader must go through entire report to understand main findings and their implications.

self-administration decreased by 50% or more in four of the five subjects (range 50 to 67%). Average cocaine self-administration decreased by 49 ± 15.5% to 1.60 ± 0.25 mg/kg per day during the first 5 days of buprenorphine treatment ($P < 0.01$). Average cocaine self-administration then decreased to 77 ± 7.4% and 83 ± 8.2% below base-line levels during buprenorphine treatment days 6 to 10 and 11 to 15, respectively. Cocaine self-administration averaged 0.98 ± 0.11 mg/kg per day over the first 15 days of buprenorphine treatment at 0.40 mg/kg per day.

. . . In contrast to its dose-dependent effects on cocaine self-administration, buprenorphine administration (0.40 mg/kg per day) suppressed food-maintained responding by 31 ± 8.3% during the first 15 days of treatment. Then food self-administration gradually recovered to average 20 ± 12.5% below base line during the second 15 days of treatment with a higher dose of buprenorphine. Although these changes were statistically significant ($P < 0.05$ to 0.01), it is unlikely that they were biologically significant. There were no correlated changes in body weight and animals continued to eat daily fruit and vegetable supplements. Moreover, food self-administration during the first daily session after buprenorphine treatment was not suppressed in comparison to saline treatment. The distribution of food intake across the four daily food sessions was equivalent during saline and buprenorphine treatment conditions. Four of five animals returned to base-line levels of food-maintained operant responding within 3 to 17 days after cessation of buprenorphine treatment (8.5 ± 2.9 days). Animals were not sedated during buprenorphine treatment and activity levels appeared normal. These data suggest that buprenorphine treatment suppressed cocaine-maintained responding but did not produce a generalized suppression of behavior.

but

Abstract: Cocaine abuse has reached epidemic proportions in the United States, and the search for an effective pharmacotherapy continues. Because primates self-administer most of the drugs abused by humans, they can be used to predict the abuse liability of new drugs and for preclinical evaluation of new pharmacotherapies for drug abuse treatment. Daily administration of buprenorphine (an opioid mixed agonist-antagonist) significantly suppressed cocaine self-administration by rhesus monkeys for 30 consecutive days. The effects of buprenorphine were dose-dependent. The suppression of cocaine self-administration by buprenorphine did not reflect a generalized suppression of behavior. These data suggest that buprenorphine would be a useful pharmacotherapy for treatment of cocaine abuse. Because buprenorphine is a safe and effective pharmacotherapy for heroin dependence, buprenorphine treatment may also attenuate dual abuse of cocaine and heroin.

Solution: Editor provides abstract* of report giving objective, findings, and significance.

*Terms marked with an asterisk are defined in the Dictionary.

Cocaine abuse is widespread in the general population and has also increased among heroin-dependent persons, including those in methadone maintenance treatment programs. The many adverse medical consequences of cocaine abuse are augmented by the combined use of cocaine and heroin. For example, dual addiction to intravenous cocaine and heroin may increase the risk of acquired immunodeficiency syndrome (AIDS), both through needle sharing and through the combined immunosuppressive effects of both drugs. Intravenous drug abuse was estimated to account for more than 30% of AIDS victims in the United States in 1988.

At present, there is no uniformly effective pharmacotherapy for cocaine abuse, and the dual abuse of cocaine plus heroin is an even more difficult treatment challenge . . . [document continues with full report][1]

IN LONGER DOCUMENTS, TRY STATEMENTS OF MAIN IDEAS AND SIGNIFICANCE AT THE TOP OF THE DOCUMENT, AT THE TOP OF COMPLICATED SECTIONS, AND BEFORE ANY COMPLICATED PARAGRAPH OR SERIES OF PARAGRAPHS.

Try a combination of prefatory sentences that (1) summarize main points, (2) show division of the argument, and (3) show the significance of the details that follows.

not

Demographic Background

Major demographic and sociological changes directly influencing family composition have taken place in this century, with the pace of change accelerating in the past two decades.

A decline in fertility rates was, in many cases, interrupted by the post-World War II baby boom, but it resumed in the 1960s. Japan is an exception, in that fertility rates have declined sharply and almost continuously since the late 1940s, with no postwar upturn apart from a small recovery and stabilization from the mid-1960s to the early 1970s. The change in total fertility rates in 10 countries is shown in Table 1.

With the exception of some baby "boomlets" in the late 1970s and 1980s, total fertility rates in most developed countries have declined to below the level needed to replace population deaths, namely, 2.1 children per woman. This means that the current population will not even replace itself if current levels of fertility continue. By 1988, fertility rates in the developed countries fell into a narrow range from 1.3 to 1.4 children per woman in Germany and Italy to around 1.9 to 2.0 in the United States and Sweden.

Decreased fertility has important implications for the family. In particular, family size is getting smaller, with consequences for parents—especially mothers—and children. Probably the most significant effect of falling fertility is the opportunity it

Problem: Reader must scan entire section (3 pages) to realize that the discussion comprises four specific demographic changes.

Not readily apparent that the author has begun discussing the first division, fertility rates. Also, main point of subsection on fertility not stated at onset.

has afforded women for increased participation in the labor market. And the converse relation holds as well: increased participation leads to lower fertility. Smaller families also mean fewer relatives to care for young children.

It is important to consider the age structure of the population because different arrays of persons by age result in different household structures across countries. Mortality, as well as fertility, has declined in the 20th century. The decline in mortality has been more or less continuous, and the average age of death has risen considerably in all developed countries. . . .

Not readily apparent that author has begun discussing second subsection, "Aging of the population."

Life expectancy at birth is higher for women than for men in all the countries studied. Women outlived men by 6 to 7 years, on average, and this influences household structures, as many more women than men live alone at older ages. In most developed countries, women must anticipate a period of living alone at some point during their later years.

Aging of the population is common to all the industrialized countries, although there are considerable differences in the extent and timing of the phenomenon. These differences are reflected in the comparisons presented later on household type. For example, countries with high proportions of elderly people tend to have higher proportions of single-person households, because the elderly are increasingly living alone.

Almost everyone in the United States gets married at some time in his or her life. The United States has long had one of the highest marriage rates in the world, and even in recent years it has maintained a relatively high rate. For the cohort born in 1945, for example, 95 percent of the men have married, compared with 75 percent in Sweden. The other countries studied ranked somewhere between these two extremes.

Third division of argument not readily apparent.

In Scandinavia and Germany, the downward trend in the marriage rate was already evident in the 1960s; in the United States, Canada, Japan, France, the Netherlands, and the United Kingdom, the decline began in the 1970s. (Table 3)

Details of Table 3 given without prefatory statement summing up details.

In Europe, the average age at marriage fell until the beginning of the 1970s, when a complete reversal occurred. Postponement of marriage by the young is now common throughout the continent. The generation born in the early 1950s initiated this new behavior, characterized by both later and less frequent marriage. Average age at first marriage has also been rising in the United States since the mid-1950s, but Americans still tend to marry earlier than their European counterparts. . . .

According to Table 4, close to half of all live births in Sweden are now outside of wedlock, up from only 1 in 10 in 1960. Denmark is not far behind. In the United States, France, and the United Kingdom, unmarried women account for more than 1 out of 5 births, while the rates are far lower in the Netherlands, Italy, and Germany.

Fourth division in argument not readily apparent. No summary statement of details subsequently recounted in Table 4.

. . . A relatively high proportion of births out of wedlock in the United States and the United Kingdom are to teenagers— more than 33 and 29 percent, respectively. In Sweden, teenagers account for only 6 percent, and in France and Japan about

10 percent. More than half of the births out of wedlock in Sweden are to women between the ages of 25 and 34, while only one-quarter are to women in that age group in the United States and the United Kingdom.

All of the foregoing demographic trends have had an impact on household size and composition in the developed nations. This impact can be seen clearly in developments since 1960. [Tables not shown in this example.]

but

Demographic Background

Major demographic and sociological changes directly influencing family composition have taken place in this century, with the pace of change accelerating in the past two decades. Almost all developed countries have seen changes of four principal types: a decline in fertility rates, the aging of the population, an erosion of the institution of marriage, and a rapid increase in childbirths out of wedlock. Each of these four trends has played a part in the transformation of the modern family.

Fertility rates. Over the past century, women in industrialized countries have moved to having fewer children—that is, to lower fertility rates. This decline was, in many cases, interrupted by the post-World War II baby boom, but it resumed in the 1960s. Japan is an exception, in that fertility rates have declined sharply and almost continuously since the late 1940s, with no postwar upturn apart from a small recovery and stabilization from the mid-1960s to the early 1970s. The change in total fertility rates in 10 countries is shown in Table 1.

With the exception of some baby "boomlets" in the late 1970s and 1980s, total fertility rates in most developed countries have declined to below the level needed to replace population deaths, namely, 2.1 children per woman. This means that the current population will not even replace itself if current levels of fertility continue. By 1988, fertility rates in the developed countries fell into a narrow range from 1.3 to 1.4 children per woman in Germany and Italy to around 1.9 to 2.0 in the United States and Sweden.

Decreased fertility has important implications for the family. In particular, family size is getting smaller, with consequences for parents—especially mothers—and children. Probably the most significant effect of falling fertility is the opportunity it has afforded women for increased participation in the labor market. And the converse relation holds as well: increased participation leads to lower fertility. Smaller families also mean fewer relatives to care for young children.

Aging of the population. It is important to consider the age structure of the population because different arrays of persons by age result in different household structures across countries.

Solution: Editor adds summary of four main parts and their significance: "Almost all developed countries have seen changes of four principal types. . . ."

Editor adds subhead, *Fertility rates,* and inserts statement summarizing main point of details on fertility that follows: "Over the past century, women in industrialized countries have moved to having fewer children—that is, to lower fertility rates."

Editor adds second subhead: *Aging of the population.*

Mortality, as well as fertility, has declined in the 20th century. The decline in mortality has been more or less continuous, and the average age of death has risen considerably in all developed countries. . . .

Life expectancy at birth is higher for women than for men in all the countries studied. Women outlive men by 6 to 7 years, on average, and this influences household structures, as many more women than men live alone at older ages. In most developed countries, women must anticipate a period of living alone at some point during their later years.

Aging of the population is common to all the industrialized countries, although there are considerable differences in the extent and timing of the phenomenon. These differences are reflected in the comparisons presented later on household type. For example, countries with high proportions of elderly people tend to have higher proportions of single-person households, because the elderly are increasingly living alone.

Marriage and divorce. Almost everyone in the United States gets married at some time in his or her life. The United States has long had one of the highest marriage rates in the world, and even in recent years it has maintained a relatively high rate. For the cohort born in 1945, for example, 95 percent of the men have married, compared with 75 percent in Sweden. The other countries studied ranked somewhere between these two extremes.

Editor adds third subhead: *Marriage and divorce.*

According to Table 3, a trend toward fewer marriages is plain in all of the countries studied, although the timing of this decline differs from country to country. In Scandinavia and Germany, the downward trend in the marriage rate was already evident in the 1960s; in the United States, Canada, Japan, France, the Netherlands, and the United Kingdom, the decline began in the 1970s.

Editor summarizes point in Table 3 that details will illustrate.

In Europe, the average age at marriage fell until the beginning of the 1970s, when a complete reversal occurred. Postponement of marriage by the young is now common throughout the continent. The generation born in the early 1950s initiated this new behavior, characterized by both later and less frequent marriage. Average age at first marriage has also been rising in the United States since the mid-1950s, but Americans still tend to marry earlier than their European counterparts. . . .

Births out of wedlock. Rates of births to unmarried women have increased in all developed countries except Japan. (See Table 4.) Close to half of all live births in Sweden are now outside of wedlock, up from only 1 in 10 in 1960. Denmark is not far behind. In the United States, France, and the United Kingdom, unmarried women account for more than 1 out of 5 births, while the rates are far lower in the Netherlands, Italy, and Germany.

Editor adds subhead, *Births out of wedlock;*, adds summary of details in Table 4. ("Rates of births to unmarried women have increased in all developed countries except Japan.")

. . . A relatively high proportion of births out of wedlock in the United States and the United Kingdom are to teenagers—

more than 33 and 29 percent, respectively. In Sweden, teen-agers account for only 6 percent, and in France and Japan about 10 percent. More than half of the births out of wedlock in Sweden are to women between the ages of 25 and 34, while only one-quarter are to women in that age group in the United States and the United Kingdom.

All of the foregoing demographic trends have had an impact on household size and composition in the developed nations. This impact can be seen clearly in developments since 1960.[2]

FOCUS VAGUE LANGUAGE.

not

This is an interesting new method

but

The British Library has developed a new, low-cost method to preserve books. The method, which uses gamma rays to irradiate chemically treated paper, costs only $8 per volume—a significant savings over the $340 price tag for the present method.

Problem: Sentence too general; doesn't address question of how or why the method is interesting.

Solution: Editor provides opening sentences that summarize ensuing information.

not

Abstract: This paper discusses how buprenorphine suppresses cocaine self-administration by rhesus monkeys. Results are presented and conclusions drawn. Finally, implications are discussed in terms of the general problem of drug abuse.

but

Abstract: Daily administration of buprenorphine significantly suppressed cocaine self-administration by rhesus monkeys for 30 consecutive days. The effects of buprenorphine were dose-dependent. These data suggest that buprenorphine would be a useful pharmacotherapy for treatment of cocaine abuse.

Problem: Summary vague; says "results are presented and conclusions drawn," but does not tell actual results or conclusions.

Solution: Editor makes language more specific; gives actual results, conclusions, and implications.

TRIM PROCEDURAL DETAIL IN SUMMARIES OR ABSTRACTS TO LEAVE ROOM FOR MAIN POINTS.

Use summaries to tell readers main points and why these points matter.

not

Abstract: We studied 3012 patients who were hospitalized at the University of St. Clair, New Jersey, in 1976, 1977, and 1987. Ten diagnoses were selected, including four medical diagnoses. The methods used to update 1976 and 1977 prices to 1987 prices were similar to those used by Sinclair, Stone, and others. (*1*) The medical records of each subject were reviewed to confirm the diagnosis and to collect key demographic and clinical data. Services used by patients were compared in constant 1987 prices. Included were professional fees that

Problem: The procedural information is far too detailed, taking up space needed for the objective and the findings.

were billed by the hospital, such as fees for radiology, pathology, and cardiology. The study provided strong evidence that for most, but not all, of the 10 diagnoses studied at the University, changes in styles of treatment between 1976 and 1987 contributed little to higher hospital cost. The study found that procedures such as laboratory tests contributed little to rising costs, and that new imaging techniques were commonly substituted for older procedures. The primary clinical factors associated with rising costs, we concluded, were the provision of surgery to patients admitted for acute myocardial infarction, rds of the newborn, delivery, and other intensive treatments for the critically ill.

but

Abstract: To assess whether changes in clinical practices have contributed to rising hospital costs, we studied 3012 patients who were hospitalized at the University of St. Clair, New Jersey, in 1976, 1977, and 1987. For most of the 10 diagnoses studied, there was little change in total use of services by patients. The length of stay and the use of laboratory services were either the same or in decline. Only for patients with acute myocardial infarction did the use of imaging procedures increase substantially. Procedures such as laboratory tests did not contribute to rising costs, and the new imaging techniques were commonly substituted for older, more invasive procedures. The primary causes of rising costs were the provision of surgery to patients admitted for acute myocardial infarction, delivery, respiratory distress syndrome of the newborn, and the provision of other intensive treatments for the critically ill.[3]

Solution: Editor adds statement of objective; shortens procedural details; deletes expendable language. ("The study found that. . . . we concluded. . . . ")

Editor deletes extraneous reference; defines *rds* ("respiratory distress syndrome").

2/////

EDIT FOR TITLES, HEADINGS, AND SUBHEADINGS THAT MARK MAIN POINTS AND THEIR ORDER

Titles,* headings,* and displayed lists* or text allow readers to see main points, sort major and minor categories, and scan documents effectively. They help readers process technical information more quickly.

This chapter discusses some ways editors intervene to make titles, headings, and displayed lists and text informative, distinctive, and consistent.

USE THE TITLE TO GIVE THE OBJECTIVE.

Readers use titles or subject lines to preview documents. Make sure the title states the objective and, insofar as possible, why this objective matters. Many writers understate in their titles.

not

Notes on a New Operation: Selective Posterior Rhizotomy

but

Selective Posterior Rhizotomy: Neurosurgical technique to reduce spasticity in youths with cerebral palsy.

Problem: Title unfocused, does not suggest range or limits of technique, or whether technique is effective.

Solution: Editor provides title that gives objective and significance of work.

CAST THE TITLE IN THE ACTIVE VOICE* TO EMPHASIZE THE SUBJECT/AGENT, IN THE PASSIVE TO EMPHASIZE THE OBJECT.

Title emphasizes subject/agent:
Buprenorphine Suppresses Cocaine Self-Administration in Rhesus Monkeys

Title emphasizes object:
The Suppression of Cocaine Self-Administration in Rhesus Monkeys by Buprenorphine

*Terms marked with an asterisk are defined in the Dictionary.

USE HEADINGS AND SUBHEADINGS TO MARK MAIN POINTS AND MAJOR DIVISIONS.

not

Evidence that diet is one factor in cancer comes from two entirely different kinds of data: laboratory analyses and population studies.

Many foods contain substances that may influence cancer formation in humans. Natural carcinogens are found in a number of foods, including potatoes, parsley, and celery. Processing may add potential carcinogens to other foods. For example, the nitrates that are used to preserve meats such as bacon can combine with amines in the body to form nitrosamines, which are known carcinogens. Charcoal broiling causes the formation of carcinogens called polycyclic hydrocarbons in meat.

Another source is population studies comparing the incidence of specific cancers in populations with different dietary patterns; these have also yielded valuable data. For example, the Japanese diet contains much less fat than the American diet, and the Japanese have a much lower incidence of breast cancer and colon cancer than Americans. This kind of evidence is suggestive but not conclusive. It is possible that some other element of the diet that is closely associated with fat is the causative agent, or that the overall diet, rather than individual nutrients, is responsible for the increased cancer risk.

but

Two Kinds of Evidence

Evidence that diet is one factor in cancer comes from two entirely different kinds of data: laboratory analyses and population studies.

Laboratory analyses. Many foods contain substances that may influence cancer formation in humans. Natural carcinogens are found in a number of foods, including potatoes, parsley, and celery. Processing may add potential carcinogens to other foods. For example, the nitrates that are used to preserve meats such as bacon can combine with amines in the body to form nitrosamines, which are known carcinogens. Charcoal broiling causes the formation of carcinogens called polycyclic hydrocarbons in meat.

Population studies. Studies that compare the incidence of specific cancers in populations with different dietary patterns have also yielded valuable data. For example, the Japanese diet contains much less fat than the American diet, and the Japanese have a much lower incidence of breast cancer and colon cancer than Americans. This kind of evidence is suggestive but not

Problem: No headings or subheadings to mark divisions for the reader.

Solution: Editor adds boldface* section heading (**Two Kinds of Evidence**).

Editor adds italicized* subheadings (*Laboratory analyses . . . Population studies*) to show reader the division of information.

*Terms marked with an asterisk are defined in the Dictionary.

conclusive. It is possible that some other element of the diet that is closely associated with fat is the causative agent, or that the overall diet, rather than individual nutrients, is responsible for the increased cancer risk.

USE TITLES, HEADINGS, AND SUBHEADINGS TO GROUP TASKS OR LONG LISTS.

not

INSTRUCTIONS

1. Hold screen vertically with legs down.
2. Pull one leg outward. All legs will open fully at once.
3. Set screen on floor.
4. Press release tab and raise extension rod.
5. Swing case to horizontal position and place screen hanger on hook.
6. Release handle latch and raise case until it locks into place.
7. To lower screen, grasp extension rod firmly with one hand.
8. With other hand, press release tab and gently drop rod to its lowest position.
9. Unhook the screen.
10. Swing case to vertical position and snap hook back into slot.
11. Holding the screen with one hand, raise it slightly and close tripod legs.

Problem: Uninformative title; no grouping to show reader major divisions of steps.

but

How To Use the Portable Screen

To Set Up the Screen

1. Hold screen vertically with legs down.
2. Pull one leg outward. All legs will open fully at once.
3. Set screen on floor.
4. Press release tab and raise extension rod.
5. Swing case to horizontal position and place screen hanger on hook.
6. Release handle latch and raise case until it locks into place.

To Close the Screen

7. Grasp extension rod firmly with one hand.
8. With other hand, press release tab and gently drop rod to its lowest position.
9. Unhook the screen.
10. Swing case to vertical position and snap hook back into slot.

Solution: Editor adds informative title, groups two sets of steps with subheadings (**To Set Up the Screen, To Close the Screen**).

11. Holding the screen with one hand, raise it slightly and close tripod legs.

KEEP HEADINGS AND SUBHEADINGS BRIEF, INFORMATIVE, AND PARALLEL*.

not

Here are some suggestions for keeping meat, poultry, eggs, milk, cheese, and other perishable foods cold:

- Pick up the perishables as your last stop in the grocery, and get them home and into the refrigerator quickly, especially in hot weather.
- **Refrigerating** Since repeated handling can introduce bacteria to meat and poultry, leave products in the store wrap unless the wrap is damaged.
- **To Freeze** While "freezer-burn"—white, dried-out patches on the surface of meat—won't make you sick, it does make meat tough and tasteless. To avoid it, wrap freezer items in heavy freezer paper, plastic wrap, or aluminum foil. Place new items to the rear of the freezer, and old items to the front so that they'll be used first.
- **For Thawing.** To be safe, take meat or poultry out of the freezer and leave overnight in the refrigerator.

Problem: Author's subheadings are not parallel.* The first displayed item—the item after the bullet*—has no subheading at all. It begins with a sentence, "Pick up the perishables. . . ." The three following subheadings ("Refrigerating," "To Freeze," and "For Thawing") do not match grammatically. "Refrigerating" is not parallel to the infinitive "To Freeze" or the prepositional phrase "For Thawing."

but

Here are some suggestions for keeping meat, poultry, eggs, milk, cheese, and other perishable foods cold:

- **Shopping** Pick up the perishables as your last stop in the grocery, and get them home and into the refrigerator quickly, especially in hot weather.
- **Refrigerating** Since repeated handling can introduce bacteria to meat and poultry, leave products in the store wrap unless the wrap is damaged.
- **Freezing** While "freezer-burn"—white, dried-out patches on the surface of meat—won't make you sick, it does make meat tough and tasteless. To avoid it, wrap freezer items in heavy freezer paper, plastic wrap, or aluminum foil. Place new items to the rear of the freezer, and old items to the front so that they'll be used first.
- **Thawing** To be safe, take meat or poultry out of the freezer and leave overnight in the refrigerator.

Solution: Editor revises subheadings in displayed list so that they are parallel.

COMBINE EXPLICIT INTRODUCTORY STATEMENTS OF MAIN IDEAS WITH HEADINGS AND SUBHEADINGS TO MARK THE PATH OF THE ARGUMENT.

not

The next topic is how to keep food cold.

Pick up perishables as your last stop in the grocery and get

Problem: Section headings missing, information placed solely in transitional sentence, "The next topic is how to keep food cold."

*Terms marked with an asterisk are defined in the Dictionary.

them home and into the refrigerator quickly, especially in hot weather.

Since repeated handling can introduce bacteria to meat and poultry, leave products in the store wrap unless the wrap is damaged.

While "freezer burn"—white, dried-out patches on the surface of meat—won't make you sick, it does make meat tough and tasteless. To avoid it, wrap freezer items in heavy freezer paper, plastic wrap, or aluminum foil. Place new items to the rear of the freezer and old items to the front so that they'll be used first. Dating freezer packages also tells you what to use first.

The safest way to thaw meat or poultry is to take it out of the freezer and leave it overnight in the refrigerator. Normally, it will be ready to use the next day.

For faster thawing, put the frozen package in a watertight plastic bag under cold water. Change the water often. The cold water temperature slows bacteria that might grow in the outer, thawed portions of the meat while the inner areas are still thawing.

If you have a microwave oven, you can safely thaw meat and poultry in it. Follow the manufacturer's directions.

but

Keep Food Cold

The colder food is kept, the less chance bacteria have to grow. To make sure your refrigerator and freezer are giving you good protection against bacterial growth, check them with an appliance thermometer. The refrigerator should register 40°F or lower. The freezer should read 0°F or lower.

Here are some tips for keeping meat, poultry, eggs, milk, cheese, and other perishable foods cold:

- **Shopping** Don't cool leftovers on the kitchen counter. Put them straight into the refrigerator. Divide large meat, macaroni, or potato salads and large bowls of mashed potatoes or dressing into smaller portions. Food in small portions cools more quickly to temperatures where bacteria quit growing.
- **Refrigerating** Since repeated handling can introduce bacteria to meat and poultry, leave products in the store wrap unless the wrap is damaged.
- **Freezing** While "freezer burn"—white, dried-out patches on the surface of meat—won't make you sick, it does make meat tough and tasteless. To avoid it, wrap freezer items in heavy freezer paper, plastic wrap, or aluminum foil. Place new items to the rear of the freezer and old items to the front so that they'll be used first. Dating freezer packages also tells you what to use first.
- **Thawing** To be safe, take meat or poultry out of the freezer and leave it overnight in the refrigerator. Normally, it will be ready to use the next day.

For faster thawing, put the frozen package in a watertight plastic bag under cold water. Change the water often. The

---

Writer plunges into detail ("Pick up perishables . . . ") without giving adequate opening statement of main idea and significance.

Writer gives no roadmap or listing sentence to mark division of the text into parts.

Writer uses no subheadings to mark parts of list.

Solution: Editor uses section heading ("Keep Food Cold") instead of "The next topic is . . . ". Editor adds introductory paragraph of section to give main idea, significance, and supporting detail. Editor adds roadmap or listing sentence ("Here are some tips . . . ") to introduce list. Editor summarizes major items in list with boldface paragraph headings (**Shopping, Refrigerating, Freezing, Thawing**).

cold water temperature slows bacteria that might grow in the outer, thawed portions of the meat while the inner areas are still thawing.[4]

MAKE SURE THE FIRST SENTENCE FOLLOWING THE HEADING IS COMPLETE.

Do not use a pronoun in the first sentence that refers to the heading.

not

Health Consequences of Fraud

These are very serious, according to presenters at the conference. There are several consequences.

Failure to Seek Legitimate Medical Care. This is the main problem. Public health and safety can be jeopardized by false promises that divert or deter individuals from pursuing sound forms of medical treatment or that encourage them to abandon beneficial therapy for a disease. Fraud may encourage people to reject legitimate medical advice and to practice inappropriate self-medication that is less likely to be helpful.

The Food and Drug Administration's annual reports document numerous instances of fraud-induced failure to obtain appropriate health care. Because early detection and treatment improve the prognosis for many illnesses, unproven "nutritional" therapies may unnecessarily delay beneficial intervention.

Partially Toxic Food Components. These can lead to public injury. Just because a substance occurs naturally in food does not mean that it is necessarily safe. Many of the chemicals known to be present in herbs, for example, have never been tested for safety. Some plant foods contain potentially unsafe, pharmacologically active ingredients such as aflatoxin, one of the most potent carcinogens known. Few buyers are aware of the harmful components in these products.

but

Health Consequences of Fraud

According to presenters at the conference, nutrition fraud may lead to deleterious health consequences caused by

- failure to seek legitimate medical care
- potentially toxic components of foods and products
- nutrient toxicities and deficiencies.

Failure to Seek Legitimate Medical Care. Public health and safety can be jeopardized by false promises that divert or deter individuals from pursuing sound forms of medical treatment or that encourage them to abandon beneficial therapy for a dis-

Problem: First sentence after section heading is incomplete. The pronoun "These" refers to the heading.

The first sentence after the subheading is incomplete. The pronoun "This" refers to the heading.

The first sentence after the subheading is incomplete. The pronoun "These" refers to the heading.

Corrections: Editor revises so that sentences following headings are complete.

ease. Fraud may encourage people to reject legitimate medical advice and to practice inappropriate self-medication that is less likely to be helpful, and more likely to be directly harmful, than medical technology based on a sound understanding of human biology and nutrition.

The Food and Drug Administration's annual reports document numerous instances of fraud-induced failure to obtain appropriate health care. Because early detection and treatment improve the prognosis for many illnesses, unproven "nutritional" therapies may unnecessarily delay beneficial intervention. Some diet regimens recommended by health faddists to treat cancer, for example, are so nutritionally deficient or toxic that adherence to them has caused death or serious illness.

Partially Toxic Food Components. Public injury can occur when foods and unproven remedies are toxic. Just because a substance occurs naturally in food does not mean that it is necessarily safe. Many of the chemicals known to be present in herbs, for example, have never been tested for safety. Some plant foods contain potentially unsafe, pharmacologically active ingredients such as aflatoxin, one of the most potent carcinogens known. Few buyers are aware of the harmful components in these products.[5]

USE LISTING TO GIVE EACH STEP A DISTINCT VISUAL IDENTITY.

not

To repair cracks in concrete, chisel out the crack, widening it under the surface. Then clean the concrete surface thoroughly with a wire brush. Then mix a batch of mortar according to the directions on the package. Mix in the epoxy concrete with the mortar according to the directions on the epoxy container. Put the mixture into crack, using the trowel.

Problem: Order of steps not visually distinct.

but

To repair cracks in concrete,

1. Chisel out the crack, widening it under the surface.
2. Clean the concrete surface thoroughly with a wire brush.
3. Mix a batch of mortar according to the directions on the package. Mix in the epoxy concrete with the mortar according to the directions on the epoxy container.
4. Put the mixture into the crack, using the trowel.

Solution: Editor separates steps by numbers, by space between items, and by hanging paragraphs*.

MAKE CAUTIONS,* WARNINGS, AND NOTES STAND OUT.

not

Repairing Cracks in Concrete Sidewalks
Caution—Repair only when concrete is dry

Problem: Caution not visually distinct from text.

1. Chisel out the crack, widening it under the surface.
2. Clean the concrete surface thoroughly with a wire brush.
3. Mix a batch of mortar according to package directions.
4. Mix in epoxy concrete with mortar according to directions on epoxy container.
5. Put the mixture into the crack, using the trowel.

but

Repairing Cracks in Concrete Sidewalks
<u>**CAUTION: Repair Only When**</u>
<u>**Concrete Is Dry**</u>

Solution: Editor uses boldface, capitals, and underlining to make caution highly visible.

1. Chisel out the crack, widening it under the surface.
2. Clean the concrete surface thoroughly with a wire brush.
3. Mix a batch of mortar according to package directions.
4. Mix in epoxy concrete with mortar according to directions on epoxy container.
5. Put the mixture into the crack, using the trowel.

*Terms marked with an asterisk are defined in the Dictionary.

3/////

EDIT FOR DEFINITIONS* CRUCIAL TO READER UNDERSTANDING

Writers familiar with their topics often fail to define abbreviations,* acronyms, and terms; steeped in their subjects, they
assume readers are equally familiar with the shorthand of their
field. Editors confront such documents and decide when definitions should be added and when they should be expanded.
Like the addition of strong lead sentences and clear summary
paragraphs, inserted definitions are changes editors make to
documents in the hopes of enhancing readers' understanding.

When are such changes necessary? If the intended readers
are sophisticated—that is, trained and conversant in the subject—basic terms rarely require definitions. An article on a
new method of testing infants for the human immunodeficiency
virus, for instance, needs no definition of the term DNA. But
any specialized terms, or terms used with a restricted meaning,
require definitions. For instance, in the report of a new method
of testing infants for the human immunodeficiency virus, the
limits of the term "neonatal" (age less than 28 days) are essential if others are to reproduce the study and achieve comparable
results.

This section suggests typical strategies for inserting and expanding necessary definitions.

INSERT A DEFINITION WITH PARENTHESES,* DASHES,* AND EXPRESSIONS LIKE "THAT IS," "CALLED," "KNOWN AS," OR "IS DEFINED AS."

not

Information storage involves the synapses.

Problem: Author assumes readers need no definition of "synapses," even though sentence is from book-let designed for broad readership.

*Terms marked with an asterisk are defined in the Dictionary.

but

Information storage involves the synapses—the points at which information is transferred from cell to cell.

not

Amenorrhea can be caused by obesity or extreme underweight, a disorder of the hypothalamus, a deficiency of the ovaries or pituitary, or pregnancy and lactation.

but

Amenorrhea (the absence or abnormal stoppage of menstruation) can be caused by obesity or extreme underweight, a disorder of the hypothalamus, a deficiency of the ovaries or pituitary, or pregnancy and lactation.

not

Hepatic necrosis may occur.

but

Death of liver tissue, called hepatic necrosis, may occur.

Solution: Editor decides some readers may not understand term "synapses"; suggests definition, in this case inserted using dashes.

Problem: Author omits definition of "amenorrhea," although term is part of handbook for wide readership.

Solution: Editor decides some readers may not understand term "amenorrhea"; suggests insertion of definition, in this case using parentheses.

Problem: Necessary term, "hepatic necrosis," undefined.

Solution: Editor suggests insertion of definition, in this case using "called."

WRITE OUT ABBREVIATIONS AND ACRONYMS ON FIRST USE.

The abbreviation or acronym follows first use of the term and is enclosed in parentheses. The abbreviation or acronym is correct for subsequent uses.

not

SLT poses many health hazards, some so serious that several states are discussing ways to force manufacturers of SLT to provide warnings on packages and advertisements.

but

Smokeless tobacco (SLT) poses many health hazards, some so serious that several states are discussing ways to force manufacturers of SLT to provide warnings on packages and advertisements.

Problem: Author uses abbreviation, SLT, without explaining it.

Correction: Editor writes term in full on first use, followed by abbreviation in parentheses.

USE A CONSISTENT TYPE STYLE TO INTRODUCE KEY TERMS.

not

Rock materials may be classified as **consolidated** (often called bedrock) or as unconsolidated. Consolidated rock usually consists of sandstone, limestone, granite, or other rocks. Unconsolidated rock consists of granular material such as sand, gravel, and clay.

Problem: Author introduces terms inconsistently. The first term is in boldface,* the second in roman.*

*Terms marked with an asterisk are defined in the Dictionary.

but

Rock materials may be classified as *consolidated* (often called bedrock) or *unconsolidated*. Consolidated rock usually consists of sandstone, limestone, granite, or other rocks. Unconsolidated rock consists of granular material such as sand, gravel, and clay.

Correction: Editor uses same type style, in this case italic, to introduce terms.

CONSIDER AN EXAMPLE TO CLARIFY A DEFINITION.

Brief definitions may be adequate, but expand them with an example if you decide the reader needs more information to understand or appreciate an important idea.

not

The California earthquake of October 1989 measured 6.9 on the Richter scale. The 1906 San Francisco earthquake measured 7.9. The Richter scale is logarithmic. Every increase of one represents a tenfold increase in magnitude.

Problem: Author assumes definition of "logarithmic" provides enough information for readers to understand difference in magnitude of the two earthquakes. Readers, however, are not this sophisticated, and think that the difference between the two earthquakes is linear—that is, that only a difference of 1 separates the magnitude of the two earthquakes.

but

The California earthquake of October 1989 measured 6.9 on the Richter scale. The 1906 San Francisco earthquake measured 7.9. The Richter scale is logarithmic. Every increase of one represents a tenfold increase in magnitude. This means that an earthquake of 7.9 is ten times stronger than one of 6.9.

Correction: Editor adds example of term "logarithmic" (". . . an earthquake of 7.9 is ten times stronger than one of 6.9").

not

Polyvinylchloride architectural products

Problem: Author assumes term is clear to readers.

but

Polyvinylchloride architectural products such as gutters and downspouts

Correction: Editor suggests term needs examples; in this case uses "such as" to introduce them.

not

In the home, activities and appliances that spray or agitate heated water create the largest release of waterborne radon.

Problem: Author assumes term "activities and appliances that spray or agitate heated water" needs no examples.

but

In the home, activities and appliances that spray or agitate heated water (showers, dishwashers, clothes washers) create the largest release of waterborne radon.

Correction: Editor suggests examples of activities and appliances, in this case set off by parentheses, as sentence is part of broad-based pamphlet designed for U.S. homeowners.

not

There are no moving parts.

Problem: Term "moving parts" used without examples in article for popular readership.

but

There are no moving parts—no motors, no propellers, no gears, and no drive shafts.

Solution: Editor suggests adding examples of moving parts, in this case set off by a dash.

CONSIDER A COMPARISON TO CLARIFY AN EXPLANATION.

Try comparing the new term to a familiar, everyday object.

not

A pulsar emits beams of radiation. The pulsing beams are clocked to measure the rotation of the pulsar.

Problem: Author's spare explanation adequate for experts, but lacking for readers unfamiliar with pulsars.

but

A pulsar emit beams of radiation similar to those emitted by the rotating beacons on lighthouses. The pulsing beams are clocked to measure the rotation of the pulsar.

Solution: Editor suggests comparison between pulsars and familiar object, in this case "lighthouses," introduced by expression "similar to."

not

A pulsar is so dense that one with a mass equal to that of the Sun would be no more than 12 miles in diameter.

Problem: Readers need help to visualize denseness of pulsar.

but

A pulsar is so dense that one with a mass equal to that of the Sun would be no more than 12 miles in diameter. Astronomers think a single teaspoon of a pulsar would weigh 300,000 tons on Earth.

Correction: Editor suggests comparison drawn from daily experience (in this case, "a single teaspoon . . . would weigh 300,000 tons" added to illustrate density of pulsars).

not

The banjo has a fretted neck.

Problem: Readers unfamiliar with term "fretted" need a comparison.

but

The banjo has a fretted, guitar-like neck.

Solution: Editor suggests comparison, in this case using the suffix *-like*.

not

The small vapor detector. . .

Problem: Readers need example of "small" drawn from daily life.

but

The badge-sized vapor detector

Solution: Editor suggests comparison, in this case using "badge-sized."

CONSIDER AN ANALOGY,* METAPHOR, OR FIGURE* TO CLARIFY AN EXPLANATION.

When using comparisons in technical text, remember that comparisons are two-edged. They may mislead as well as illuminate. Readers may incorrectly infer that because *x* and *y* are

*Terms marked with an asterisk are defined in the Dictionary.

Fig. 2.1. *Source:* Kristin LeVier, *Science Notes,* Fall 1989.

alike in one respect, they correspond in other respects. One way to guard against this is to explain the limits of the comparison. In the following example, the passage is shown first without a necessary definition, and then with a definition, comparison, figures, and explicit acknowledgment of the limits of the comparison.

not

The conventional theory of solar system formation has established that a young solar system is a nebula. The nebula spreads and flattens as it rotates.

Problem: No definition of "nebula."

but

The conventional theory of solar system formation has established that a young solar system is a spinning cloud, or nebula, of solids and gases. The nebula spreads and flattens as it rotates. (Fig. 2-1)

First Version of Solution: Editor adds definition and figure to clarify term "nebula."

or

The conventional theory of solar system formation has established that a young solar system is a spinning cloud, or nebula, of solids and gases. The nebula spreads and flattens as it rotates, like a lump of spinning pizza dough. (Fig. 2-2)

Second Version of Solution: Editor adds comparison with pizza dough in words and figure to clarify term "nebula."

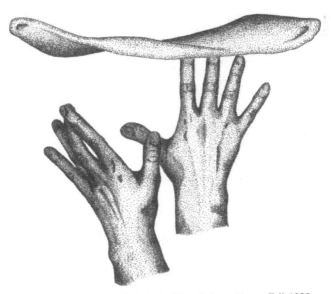

Fig. 2.2. *Source:* Kristin LeVier, *Science Notes,* Fall 1989.

or

The conventional theory of solar system formation has established that a young solar system is a spinning cloud, or nebula, of solids and gases. The nebula spreads and flattens as it rotates, like a lump of spinning pizza dough. Unlike the dough, however, the nebula isn't a solid disk, and it has a proto-sun at its center. . . .[6]

Final Version of Solution: Editor adds limits of comparison ("Unlike the dough, however . . .).

4/////

EDIT FOR A CONCISE, CONSISTENT STYLE

The term "style" has several meanings. Used in titles like "Style Guide" or "Style Manual," style is a set of rules for grammar, punctuation, and usage. Style is also the set of conventions followed by a group or publication—conventions such as whether to hyphenate "mid range" or use an apostrophe in *NMRs*. Such conventions are usually adopted for consistency, so that, for example, the three forms "workstation," "work-station," and "work station" will not appear within the same document.

No single, universally accepted book of editing standards exists. Indeed, editorial style in the sense of conventions, and even in the sense of grammar, punctuation, and usage, varies considerably among publications. What one professional society applauds in capitalization,* abbreviation,* and hyphens,* for instance, another may dislike.

The term "style" also has a third meaning. In its broadest sense, style refers to the sound of the author's voice. Here style is an amalgam of the word choices, word order, and rhythms the author employs. Authorial style in technical documents, while necessarily more muted than its literary counterpart, shows itself in such choices as use of active* or passive voice and use of first person.*

Editors intercede for style in all three of its senses. Clarity is the goal. The search for clarity leads editors to remove the static of unnecessary words, the clutter of sprawling sentences. They adjust punctuation to aid the reader's understanding, or, at the least, to eliminate some possible sources of misunderstanding. They impose standards for correctness and consistency, adjust voice to heighten or lessen emphasis. Few distinctions are too fine, for it is in such distinctions that meaning may be promoted.

The following section discusses ways to develop an eye for stylistic editing, including strategies to tighten and focus

*Terms marked with an asterisk are defined in the Dictionary.

phrases, sentences, and paragraphs, to select and strengthen voice, and to scan documents for predictable lapses in grammar, punctuation, and usage.

TRIM UNNECESSARY WORDS.*

Edit for redundancies (unnecessary repetition of words or their meanings), inflated phrases, and expendable language.

not

fused together

Problem: "Fused" means "together." Expression "fused together" is redundant.

but

fused

Solution: Editor eliminates "together."

not

performs an investigation of

Problem: Verb "investigate" buried in nominalization* "performs an investigation of."

but

investigates

Solution: Editor recovers verb "investigates" from its nominalized form "performs an investigation of."

not

The glassware was still being placed in the wrong place.

Problem: Pointless repetition of "place."

but

The glassware was still being misplaced.

Solution: Editor deletes "being placed in the wrong place"; inserts "misplaced."

TIGHTEN SENTENCES THAT SPRAWL.

Readers have difficulty with long sentences that wander. Edit to tighten and focus such sentences.

not

Each vial was originally designed by Sinclair Company to be assembled with the aid of a rubber O-ring inside the vial that subsequently formed a seal with the glass ball that was dropped after sampling to seal the vial and protect the sample from evaporation.

Problem: Author has written 45-word, sprawling sentence that demands considerable concentration from the reader.

but

Each vial was originally designed by Sinclair Company to be assembled with a rubber O-ring inside the vial. The ring held the glass ball that dropped after sampling. The ball and O-ring sealed the vial and protected the sample from evaporation.

Solution: Editor converts original sentence to three shorter sentences that present the sequence more readably.

*Terms marked with an asterisk are defined in the Dictionary.

REVISE PARAGRAPHS* THAT SPRAWL.

To follow the maxim of "One idea to a paragraph; new idea, new paragraph," focus and trim so that each paragraph is unified.

With highly detailed material, such as multi-part procedural* or results* sections, add informative headings* and subheadings to make the organizational plan clear to readers.

not

Materials and Methods

Tracheal segments were suspended in 10-mL organ baths containing Krebs-Henseleit bicarbonate buffer (g/L; NaCl 6.78, $CaCl_2$ 0.28, KCl 0.42, $MgSO_4$ 0.29, NaH_2PO_4 0.18, $NaHCO_3$ 2.1, glucose 1.0).

The buffer was maintained at 37°C and oxygenated with 95% O_2 and 5% CO_2. The tissues were attached to displacement transducers for the measurement of isometric tension. The tracheal segments were allowed to equilibrate for 60 min under a resting tension of 2 g.

For the studies of inhibition of calcium-induced contractions, the tracheal segments were exposed to 16 mmol/L KCl in normal buffer for 10 min and then washed in calcium-free buffer. The segments were then allowed to equilibrate for 40 min in calcium-free buffer and were rechallenged with 16 mmol/L KCl.

A cumulative concentration-response curve to $CaCl_2$ was then generated in the absence or presence of test compound added 20 min before the addition of $CaCL_2$.

The ability of compounds to relax respiratory smooth muscle was determined on the basal-toned trachea in normal Krebs-Henseleit bicarbonate buffer. A cumulative concentration-response curve was generated for each drug using logarithmically spaced concentrations. Each concentration of the drug was allowed to act until the relaxant response reached a stable plateau and then the next higher concentration was added. For each concentration of compound a percent relaxation was calculated.

but

Materials and Methods

Pulmonary Measurements, in vitro. Tracheal segments were suspended in 10-mL organ baths containing Krebs-Henseleit bicarbonate buffer (g/L; NaCl 6.78, $CaCl_2$ 0.28, KCl 0.42, $MgSO_4$ 0.29, NaH_2PO_4 0.18, $NaHCO_3$ 2.1, glucose 1.0). The buffer was maintained at 37°C and oxygenated with 95% O_2 and 5% CO_2. The tissues were attached to force displacement

Problem: Poorly focused paragraphs. The first two paragraphs discuss the same point; their arbitrary separation confuses readers who expect one idea to a paragraph, rather than one-half an idea to each of two paragraphs.

Material on the inhibition of calcium-induced contractions is dispersed rather than unified in a paragraph.

Solution: Editor revises paragraphs so that each is unified.

Editor adds informative headings* and subheads to show structure of section (first-level: **Materials and Methods**; second-level: *Pulmonary*

---

*Terms marked with an asterisk are defined in the Dictionary.

transducers for the measurement of isometric tension. The tracheal segments were allowed to equilibrate for 60 min under a resting tension of 2 g.

Inhibition of Calcium-Induced Contractions. For these studies, the tracheal segments were exposed to 16 mmol/L KCl in normal buffer for 10 min and then washed in calcium-free buffer. The segments were then allowed to equilibrate for 40 min in calcium-free buffer and were rechallenged with 16 mmol/L KCl. A cumulative concentration-response curve to $CaCl_2$ was then generated in the absence or presence of test compound added 20 min before the addition of $CaCL_2$.

Bronchorelaxant Activity. The ability of compounds to relax respiratory smooth muscle was determined on the basal-toned guinea pig trachea in normal Krebs-Henseleit bicarbonate buffer. A cumulative concentration-response curve was generated for each drug using logarithmically spaced concentrations. Each concentration of the drug was allowed to act until the relaxant response reached a stable plateau and then the next higher concentration was added. For each concentration of compound a percent relaxation was calculated.[7]

Measurements, in vitro; Inhibition of Calcium-Induced Contractions, and Bronchorelaxant Activity.)

AVOID NEGATIVES.

Edit for unneeded negatives. Positives are more quickly read and understood.

not

When the integrated circuit (IC) industry was in its infancy, IC manufacturers targeted the Department of Defense as their prime customer. As the industry grew, manufacturers switched from the military to the far more profitable commercial market. Within a few years the military was put in the position of not having leading-edge electronics in its purchased equipment.

Problem: Unnecessary negative ("The military was put in the position of not having leading-edge electronics") that must be translated by readers.

but

When the integrated circuit (IC) industry was in its infancy, IC manufacturers targeted the Department of Defense as their prime customer. As the industry grew, manufacturers switched from the military to the far more profitable commercial market. Within a few years, military equipment lacked leading-edge electronics.

Solution: Editor substitutes "lacked" for "put in the position of not having."

SELECT FOR VOICE.

Use the active voice* to emphasize the subject or agent. Use the passive voice to emphasize the object of the action.
    Avoid passive cliches.

To emphasize subject:
1. The committee rejected the grant.
2. The plant fabricated the rivets of soft copper, aluminum, and steel.

*Terms marked with an asterisk are defined in the Dictionary.

To emphasize object:
1. The grant was rejected by the committee.
2. The rivets were fabricated of soft copper, aluminum, and steel.

not

It has been observed that acid corrodes this substance.

but

Acid corrodes this substance.

Problem: Author uses expendable "It has been observed."

Solution: Editor deletes passive cliche.

Avoid "there was/there were" constructions when they introduce unnecessary passives and nominalizations.

not

There were expectations by the laboratory safety committee that its protocol submission would meet the deadline.

but

The laboratory safety committee expected to submit its protocols by the deadline.

Problem: "There were" introduces unnecessary nominalizations ("expectation," "submission").

Solution: Editor deletes "There were," recovers verbs ("expected," "to submit"), and changes sentence from passive to active to emphasize agent (the laboratory safety committee). Sentence more direct, far shorter.

CHECK FOR CONSISTENCY.

Many stylistic decisions depend on the set of rules the publication chooses to follow. (These are sometimes called "house rules.") Be consistent in following the rules of choice.

not

Health care plans, retirement income plans, and disability benefits protect workers and their families from financial burdens. Compensation includes a variety of employer-financed benefits, such as health care, life insurance, retirement income and paid vacation.

Problem: In the first sentence, author uses serial comma.* In the second sentence, author omits serial comma.

but

Health care plans, retirement income plans, and disability benefits protect workers and their families from financial burdens. Compensation includes a variety of employer-financed benefits, such as health care, life insurance, retirement income, and paid vacation.

Solution: Editor, who has earlier decided to use serial comma throughout document, adds serial comma in second sentence.

not

Cohabitation out of marriage became much more widespread in the 1970's, and by the late 1980s received more general acceptance in public opinion.

Problem: Author uses apostrophe* with plural of one number (1970's) but not with the other (1980s).

but

Cohabitation out of marriage became much more widespread in the 1970s, and by the late 1980s received more general acceptance in public opinion.

Solution: Editor, who has decided to omit apostrophes in plurals of numbers and capitalized abbreviations,* deletes the apostrophe in 1970s.

*Terms marked with an asterisk are defined in the Dictionary.

not

For some couples, consensual unions may be a temporary arrangement that eventually leads to marriage. For other couples consensual unions are an alternative to the institution of marriage.

but

For some couples, consensual unions may be a temporary arrangement that eventually leads to marriage. For other couples, consensual unions are an alternative to the institution of marriage.

Problem: Author sets off short introductory phrase in first sentence by a comma, but does not set off short introductory phrase in second sentence.

Solution: Editor adds comma to set off short introductory phrase in second sentence.

SCAN FOR COMMON ERRORS IN GRAMMAR, PUNCTUATION, AND USAGE: 1.

One effective way to improve your eye for stylistic lapses is to look through a group of sentences, each bearing a typical error in grammar, punctuation, or usage. If you can spot the error, go on to the next sentence. If not, review the information in a handbook or dictionary.

Accordingly, here are 79 sentences. Each has an error in punctuation, grammar, or usage. If you see more than one error, mark the more serious one.

Answers follow. Each answer includes a reference to the appropriate entries in the Dictionary that follows Part 1.

Single-Sentence Items

1. According to Dr. Sinclair, "a convention has been established to include hyperkalemia as a primary term.
2. The panel on nutrition said that considerable evidence links dietary fats with cancer in its report.
3. This is a poorly-edited manuscript.
4. He was adverse to changing the dosage
5. Each animal in the test group was isolated for six weeks before their examination.
6. The laboratory ordered a NMR spectrometer.
7. Sinclair, Lewis, and myself visited the installation.
8. Pauling's and Wilson's book on quantum mechanics is on reserve in the library.
9. Newton said, that "all bodies gravitate toward every planet."
10. Light waves are scattered by obstacles such as dust particles, gas molecules, etc.
11. Once placed in a refereed journal, scientists throughout the discipline will read the findings.
12. The principle problem with any liquid crystal display is its readability.
13. The compilation of information and its graphic representation is both emphasized.
14. The amount of personal computers increases each year.
15. Over 100 people tried the sample.

16. That he finished the job completely amazed the supervisor.
17. The assignment was given to me 7 years ago at the UCLA School of Medicine.
18. Do not enter data in this column if you have less than 10 responses.
19. A one sided sequential testing procedure was used.
20. The director of the laboratory, as well as her assistant, visit the site each week.
21. The 25 to 30-year period was chosen as the longest span for which data were available.
22. The enormity of the lasers used in the fusion experiments startled visitors to the laboratory.
23. Ten articles comprise the latest review volume on advances in laser spectroscopy.
24. It is as much, if not more, than we expected.
25. Neither the software nor the peripherals has been carefully tested.
26. Have you filed the "Request for Proposal?"
27. He did not know whether or not to measure the rainfall.
28. The consumer has a choice between hundreds of models.
29. The tenacity of the U.S. effort suggested their certainty of success.
30. The two lasers have wavelengths of 488 nm and 647 nm, respectively.
31. Dr. T. Kilson is the principle in the investigation.
32. We may categorize the ideal agent as broad spectrum, fungicidal, available for oral administration, and has no toxicity.
33. The staff submitted their demands.
34. Give the evaluations to Sinclair, Lewis, or myself.
35. The banjo is a 4, 5, or 12 stringed musical instrument.
36. The mean data is given in Table 1.
37. Many in the group found his most recent report puzzling, like its predecessors.
38. Thionyl chloride will react with the trimethyl phosphite in the next reaction, it must therefore be removed.
39. The Nobel prize system has a serious flaw; former winners nominate candidates, a procedure that makes laureates the targets of many ambitious fellow scientists.
40. The eight plastic bottles should be mailed to participants, not the glass ones.
41. The heat and mass transfer models presently composed are inadequate for dealing with the complexities of coal devolatilization, however, they are a productive beginning.
42. Antibiotics were supplied as follows; imipenem, cefoxitin, and norfloxacin (Merck Sharp & Dohme, USA), ceftizoxime (SmithKline Beecham, USA), ceftriaxone (Hoffmann-LaRoche, USA), penicillin G (Eli Lilly, USA), spectinomycin (Upjohn, USA), and tetracycline (Lederle, USA).
43. 15 mL was administered at 6:00 p.m. and 3:30 a.m.
44. Thomas Edison had one inviolate procedure in keeping his

laboratory notebooks; he made sure all entries were signed, dated, and properly witnessed.

45. Arcs are not added in the same fashion as nodes; although the accompanying manual fails to clarify the distinction.

46. Buprenorphine suppresses cocaine self administration by rhesus monkeys.

47. Cigarette smokers are involved in 40 percent more motor vehicle accidents than nonsmokers; although the association between smoking and accidents may not involve cigarettes.

48. In 1982, more than 2 million patients with burns required medical attention in the United States and more than 10,000 burn related deaths occurred.

49. The Patent & Trademark Office is revising its patent fees, considering requirements for patent applications containing DNA, RNA, and protein sequence disclosures.

50. Patients received one of four possible treatment regimens; aspirin (325 mg four times daily), sulfinpyrazone (200 mg four times daily), both, or neither.

51. Twenty seven patients participated in the study of third-degree burns that involved 50 percent of the total body surface.

52. The rank order of activity from the most to the least active agent was as follows: ceftizoxime, MIC90 = 0.015 μg/mL, ceftriaxone, MIC90 = 0.06 μg/mL, norfloxacin, MIC90 = 0.125 μg/mL, imipenem, MIC90 = 0.5 μg/mL.

53. The characteristic of ignitability applies; as liquid scintillation waste solvents have flash points below this temperature.

54. Fire related deaths in the United States decreased remarkably between 1978 and 1982.

55. Such sheets which are two to eight cells thick when applied have recently been used successfully to cover the wounds of two children with burns that affected more than 90 percent of the body surface.

56. The articles have been organized as follows: high-dosage corticosteroid treatment with cerebral malaria, 5 articles, moderate- to low-dosage corticosteroid treatment with cerebral malaria, 5 articles, unspecified dosage corticosteroid treatment with cerebral malaria, 4 articles, editorials and replies to other articles, 21 articles.

57. The effects of buprenorphine were dose dependent.

58. Because of these problems permanent skin substitutes are being designed.

59. The graphics package provides these capabilities; window-based interface, raster image display, and standalone operation.

60. Cocaine abuse is widespread in the general population and has also increased among heroin dependent persons, including those in methadone programs.

61. Students seem reluctant to appropriately attach importance to the physician's communication skills, to medical ethics, or to the circumstances of a patient's life.

62. Application of topical water-soluble antibiotics controls infection in an open wound, but they also appear to increase inflammation.
63. Structural fires account for fewer than 5 percent of hospital admissions for burns, but they are responsible for more than 45 percent of burn related deaths.
64. After gel electrophoresis and blotting a single antigenic band was revealed that comigrated with the native click beetle luciferase.
65. Although infection is the primary initiator of the state, it is now clear that the devitalized tissue itself can also initiate and perpetuate the mediator induced response.
66. Women who are thirty or older, show reduced fertility.
67. In other studies of carcinogens and tumor-promoting agents an increase in cell proliferation has accompanied increased development of tumors.
68. Previous studies with intravenous treatment have however yielded conflicting results.
69. Although unburned skin can be used up to three times for autograft donations the amount of skin still may be insufficient.
70. Reused donor split grafts with the exception of those from the scalp, have very little remaining dermis.
71. Clearly, the major advances in volume resuscitation after severe thermal injury occurred between 1964 and 1974, when the need to infuse burn patients with large quantities of salt containing fluid was identified.
72. Walsh, et. al, described a group of 11 men with thrombocytopenia.
73. Stability studies need only be done on three lots of 50-tablet bottles.
74. Percentages of the total number of abstracts for each category is given with the actual number of abstracts in parentheses.
75. To avoid mechanism based toxicity, a drug must have a high selectivity for its fungal target.
76. Because of its' poor solubility in water, parenteral administration is required.
77. There were 14 million divorced persons in the U.S. in 1988.
78. The table below shows the increasing number of office and plant workers with catastrophic medical protection during 1992.
79. Keep us informed of any plans or commitments that might effect your reading of galley proofs or page proofs during the next three months.

ANSWERS AND DICTIONARY REFERENCES

1. According to Dr. Sinclair, "a convention has been established to include hyperkalemia as a primary term."
See **Quotation marks and quotations.**
2. In its report, the panel on nutrition said that considerable evidence links dietary fats with cancer. (Equally accept-

able: The panel on nutrition said in its report that considerable evidence links dietary fats with cancer.)
See **Modifiers, misplaced or dangling.**

3. This is a poorly edited manuscript.
   See **Hyphen.**
4. He was averse to changing the dosage.
   See **Averse/adverse.**
5. Each animal in the test group was isolated for six weeks
   before its examination.
   See **Pronouns.**
6. The laboratory ordered an NMR spectrometer.
   See **A, An.**
7. Sinclair, Lewis, and I visited the installation.
   See **Pronouns.**
8. Pauling and Wilson's book on quantum mechanics is on
   reserve in the library.
   See **Possessives.**
9. Newton said that "all bodies gravitate toward every
   planet."
   See **Quotation marks and quotations.**
10. Light waves are scattered by obstacles such as dust parti-
    cles and gas molecules.
    See **etc.**
11. Scientists throughout the discipline will read the findings,
    once they are placed in a refereed journal.
    See **Modifiers, misplaced or dangling.**
12. The principal problem with any liquid crystal display is its
    readability.
    See **Principal vs. principle.**
13. The compilation of information and its graphic representa-
    tion are both emphasized.
    See **Number of subject and verb.**
14. The number of personal computers increases each year.
    See **Amount vs. number.**
15. More than 100 people tried the sample.
    See **Greater than or more than vs. over.**
16. That he completely finished the job amazed the super-
    visor. The supervisor was completely amazed that he
    finished the job. (Choice depends on intended meaning.)
    See **Modifiers, misplaced or dangling.**
17. The assignment was given to me seven years ago at the
    UCLA School of Medicine.
    See **Numbers, figures vs. words.**
18. Do not enter data in this column if you have fewer than 10
    responses.
    See **Fewer, less than.**
19. A one-sided sequential testing procedure was used.
    See **Hyphen.**
20. The director of the laboratory, as well as her assistant,
    visits the site each week.
    See **Number of subject and verb.**
21. The 25- to 30-year period was chosen as the longest span
    for which data were available.
    See **Hyphen.**

22. The enormousness of the lasers used in the fusion experiments startled visitors to the laboratory.
See **Enormity, enormousness.**

23. The latest review volume on advances in laser spectroscopy comprises ten articles.
See **Comprise.**

24. It is as much as, if not more than, we expected.
See **As much as.**

25. Neither the software nor the peripherals have been carefully tested.
See **Number of subject and verb.**

26. Have you filed the "Request for Proposal"?
See **Quotation marks and quotations.**

27. He did not know whether to measure the rainfall.
See **Whether.**

28. The consumer has a choice among hundreds of models.
See **Between vs. among.**

29. The tenacity of the U.S. effort suggested the Americans' certainty of success.
See **Pronouns.**

30. The two lasers have wavelengths of 488 nm and 647 nm each.
See **Respectively.**

31. Dr. T. Kilson is the principal in the investigation.
See **Principal vs. principle.**

32. We may categorize the ideal agent as broad spectrum, fungicidal, available for oral administration, and nontoxic.
See **Parallel construction.**

33. The staff submitted its demands.
See **Pronouns.**

34. Give the evaluations to Sinclair, Lewis, or me.
See **Pronouns.**

35. The banjo is a 4-, 5-, or 12-stringed musical instrument.
See **Hyphen.**

36. The mean data are given in Table 1.
See **Greek and Latin words and expressions.**

37. Many in the group found his most recent report, like its predecessors, puzzling.
See **Modifiers, misplaced or dangling.**

38. Thionyl chloride will react with the trimethyl phosphite in the next reaction; it must therefore be removed.
See **Semicolon.**

39. The Nobel prize system has a serious flaw: former winners nominate candidates, a procedure that makes laureates the targets of many ambitious fellow scientists.
See **Colon.**

40. The eight plastic bottles, not the glass ones, should be mailed to participants.
See **Modifiers, misplaced or dangling.**

41. The heat and mass transfer models presently composed are inadequate for dealing with the complexities of coal devolatilization; however, they are a productive beginning.
See **Semicolon.**

42. Antibiotics were supplied as follows: imipenem, cefoxitin, and norfloxacin (Merck Sharp & Dohme, USA), ceftizoxime (SmithKline Beecham, USA), ceftriaxone (Hoffmann-LaRoche, USA), penicillin G (Eli Lilly, USA), spectinomycin (Upjohn, USA), and tetracycline (Lederle, USA).
    See **Colon** and **Semicolon**.
43. Fifteen milligrams was administered at 6:00 p.m. and 3:30 a.m.
    See **Numbers, figures vs. words**.
44. Thomas Edison had one inviolate procedure in keeping his laboratory notebooks: he made sure all entries were signed, dated, and properly witnessed.
    See **Colon**.
45. Arcs are not added in the same fashion as nodes, although the accompanying manual fails to clarify the distinction.
    See **Semicolon**.
46. Buprenorphine suppresses cocaine self-administration by rhesus monkeys.
    See **Hyphen**.
47. Cigarette smokers are involved in 40 percent more motor vehicle accidents than nonsmokers, although the association between smoking and accidents may not involve cigarettes.
    See **Semicolon**.
48. In 1982, more than 2 million patients with burns required medical attention in the United States and more than 10,000 burn-related deaths occurred.
    See **Hyphen**.
49. The Patent and Trademark Office is revising its patent fees, considering requirements for patent applications containing DNA-, RNA-, and protein-sequence disclosures.
    See **Hyphen**.
50. Patients received one of four possible treatment regimens: aspirin (325 mg four times daily), sulfinpyrazone (200 mg four times daily), both, or neither.
    See **Colon**.
51. Twenty-seven patients participated in the study of third-degree burns that involved 50 percent of the total body surface.
    See **Hyphen**.
52. The rank order of activity from the most to the least active agent was as follows: ceftizoxime, MIC90 = 0.015 μg/mL; ceftriaxone, MIC90 = 0.06 μg/mL; norfloxacin, MIC90 = 0.125 μg/mL; imipenem, MIC90 = 0.5 μg/mL.
    See **Semicolon**.
53. The characteristic of ignitability applies, as liquid scintillation waste solvents have flash points below this temperature.
    See **Semicolon**.
54. Fire-related deaths in the United States decreased remarkably between 1978 and 1982.
    See **Hyphen**.

55. Such sheets, which are two to eight cells thick when applied, have recently been used successfully to cover the wounds of two children with burns that affected more than 90 percent of the body surface.
    See **Comma.**
56. The articles have been organized as follows: high-dosage corticosteroid treatment with cerebral malaria, 5 articles; moderate- to low-dosage corticosteroid treatment with cerebral malaria, 5 articles; unspecified dosage corticosteroid treatment with cerebral malaria, 4 articles; editorials and replies to other articles, 21 articles.
    See **Semicolon.**
57. The effects of buprenorphine were dose-dependent.
    See **Hyphen.**
58. Because of these problems, permanent skin substitutes are being designed.
    See **Comma.**
59. The graphics package provides these capabilities: window-based interface, raster image display, and standalone operation.
    See **Colon.**
60. Cocaine abuse is widespread in the general population and has also increased among heroin-dependent persons, including those in methadone programs.
    See **Hyphen.**
61. Students seem reluctant to attach appropriate importance to the physician's communication skills, to medical ethics, or to the circumstances of a patient's life.
    See **Infinitives.**
62. Application of topical water-soluble antibiotics controls infection in an open wound, but it also appears to increase inflammation.
    See **Pronouns.**
63. Structural fires account for fewer than 5 percent of hospital admissions for burns, but they are responsible for more than 45 percent of burn-related deaths.
    See **Hyphen.**
64. After gel electrophoresis and blotting, a single antigenic band was revealed that comigrated with the native click beetle luciferase.
    See **Comma.**
65. Although infection is the primary initiator of the state, it is now clear that the devitalized tissue itself can also initiate and perpetuate the mediator-induced response.
    See **Hyphen.**
66. Women who are thirty or older show reduced fertility.
    See **Comma.**
67. In other studies of carcinogens and tumor-promoting agents, an increase in cell proliferation has accompanied increased development of tumors.
    See **Comma.**
68. Previous studies with intravenous treatment have, however, yielded conflicting results.
    See **Comma.**

69. Although unburned skin can be used up to three times for autograft donations, the amount of skin still may be insufficient.
    See **Comma.**
70. Reused donor split grafts, with the exception of those from the scalp, have very little remaining dermis.
    See **Comma.**
71. Clearly, the major advances in volume resuscitation after severe thermal injury occurred between 1964 and 1974, when the need to infuse burn patients with large quantities of salt-containing fluid was identified.
    See **Hyphen.**
72. Walsh et al. described a group of 11 men with thrombocytopenia.
    See **et al.**
73. Stability studies need be done only on three lots of 50-tablet bottles.
    See **Only, placement of.**
74. Percentages of the total number of abstracts for each category are given with the actual number of abstracts in parentheses.
    See **Number of subject and verb.**
75. To avoid mechanism-based toxicity, a drug must have a high selectivity for its fungal target.
    See **Hyphen.**
76. Because of its poor solubility in water, parenteral administration is required.
    See **Pronouns.**
77. There were 14 million divorced persons in the United States in 1988.
    See **Abbreviations and acronyms.**
78. Table 1 shows the increasing number of office and plant workers with catastrophic medical protection during 1992.
    See **Above, below.**
79. Keep us informed of any plans or commitments that might affect your reading of galley proofs or page proofs during the next three months.
    See **Affect, effect.**

SCAN FOR COMMON ERRORS IN GRAMMAR, PUNCTUATION, AND USAGE: 2.

Another way to improve your eye for stylistic lapses is to look through sentences grouped by type of error, see if you can spot and correct the error, and, when necessary, read the related dictionary or handbook entry.

Accordingly, here is a list of sentences and paragraphs grouped by type of error. If you see more than one error, mark the more serious one. Use the Dictionary that follows these exercises for reference.

1.

1. Analysis of the tracer readily showed that "X" and "Y" impurities were present in the uv, they were well separated from the other components, and easily quantified.
2. The reaction mixture was quenched with 90%-saturated NH4Cl, the layers separated, and the solution was concentrated.
3. Protocol
   a. Resuspend the samples in acetonitrile (3.5 mL for XYZ samples and 2 mL for XYQ samples).
   b. Sonicate for 5 min.
   c. To XYZ samples, add 6.5 mL water. For XYQ samples, add 8 mL water.
   d. Wash samples with 10 mL acetonitrile: water (35:65 for XYZ-containing samples and 20:80 for XYQ-containing samples).
   e. Samples are resuspended in ethanol and sonicated for 5 min.
   f. Keep samples in capped microtubes at $-20°$ C until assayed.
4. The new method:
   • Strengthens the wood fibers of the pages by connecting them with polymer bridges.
   • Polymer bridges are created by placing the books in an alkaline monomer mixture in an oxygen-free environment.
   • No residual radiation remains in the book at the completion of the process.
   • The pages, while significantly strengthened, have the same "feel" as the original.
5. This study addressed the following advantages of buprenorphine pharmacotherapy:
   • It is better suited to outpatient therapy.
   • Able to use with codependency such as cocaine and heroin.
   • Substitute addiction is not a problem.

2.

6. A Swan-Ganz catheter was entered in the jugular vein and was advanced until the catheter tip is in or near the right atrium.
7. After the initial observation was made that transcription could be induced in quiescent cells by treating them with an appropriate growth factor, numerous follow-up studies have attempted to define the role of this transforming gene in the regulation of the mammalian cell cycle.
8. To accommodate both of these aims, the work described in the first section of this report sought to provide a general examination of reactions of oxygenated 9,10-anthraquinones with thionyl chloride. The section deals with the

synthesis of two of the viocristins and also described the preparation of the highly substituted 9,10-anthraquinones.

3.

9. 4 broths were tested (2 batches from culture X and 2 from culture Y).
10. You must give fifteen days' notice before using the equipment.
11. 6.5 mL water was added to the samples.
12. So far we have completed 5 trials.
13. We did a thirty-hour run on the test machine.
14. Table Two describes the major demographic and pregnancy-related characteristics used in the analyses, and Table Three describes the occupational exposures of the nurses during the study pregnancy.
15. After the initial analysis was completed, the variable for "all anti-neoplastic drugs" was removed from the analysis, and individual drugs reported by at least ten women were tested one at a time.
16. Many nurses were exposed to several of these drugs during the first trimester (Fig. One), and the likelihood of exposure to more than one was high.

4.

17. The report, "AIDS-related Complex in a Heterosexual Man Seven Weeks after a Transfusion," appears in the November 7 issue of *The New England Journal of Medicine*.
18. The issue also contained the report, "Failure of Extracranial-intracranial arterial Bypass to Reduce the Risk of Ischemic Stroke: Results of an International Randomized Trial."
19. The total-ion-current chromatograms obtained through gas chromatography-mass spectrometry for the analysis of steroid monosulfates and disulfates in plasma from the patient and a normal three-year-old boy are shown in figure 2 and figure 4.
20. Quantitative measurements of urinary steroids are given in table 3.

5.

21. Add 6.5 mLs water to the sample.
22. The spectrums of white petrolatum, arlacel, and octyldodecanol follow in Attachment 1.
23. The largest tube size (60 gms) was chosen.

6.

24. The breakdown of each individual's time within the division (Attachment A) is attached in alphabetical order by department.
25. The reason why charges to the departments are so large is because people are making a mistake in entering these charges.
26. The meeting will begin at 12:00 noon. Please make every effort to be there on time.
27. Besides personal enrichment, the experience of this course should enable me to assist the supervisor with the computerization of appropriate functions.
28. In the previous report, it was reported that the parameters TCA, CA+, and nCa were analyzed.
29. Urine, feces, expired air, and cage rinse were sampled at 24-hour time intervals over a period of 96 hours.
30. The purpose of this document is to report the level of two inflammatory mediators thought to be involved.
31. It was concluded that the Process Division should conduct an investigation of the matter as soon as conceivably possible.
32. After one month, the system will be reviewed and, if needed, be reassessed so that we may improve upon the system to accommodate individual needs.
33. Because this will have an effect upon our staff hiring, Ms. Sinclair should make an assessment of the funds available to us and report back to us on the matter.
34. Prior to the conference we will have a meeting to discuss what we should do in reaction to these events.
35. It was a contributing factor in the experiment.
36. We examined the patient, a three-year-old-girl, for a skin rash.
37. The matter has been referred back to the lab committee.
38. He chose a template that was rectangular in shape.
39. In order to analyze the data we have hired a statistician.
40. If the drug dosage is of a solid nature (e.g., tablets, capsules), then stability studies are done for the largest and smallest containers.

ANSWERS

Group 1: See **Parallel construction.**
Group 2: See **Tense of verbs.**
Group 3: See **Numbers, figures vs. words.**
Group 4: See **Capitalization.**
Group 5: See **Plurals.**
Group 6: See **Unnecessary words.**

5/////

EDIT FOR TABLES* THAT DISPLAY DATA VIVIDLY AND CONCISELY

USE INFORMATIVE, VISUALLY DISTINCT TITLES.*

Titles should epitomize main ideas or major benefits. Set off titles with larger type, boldface, spacing, or shading.

not

Table 3 Some Complications

Deaths
  • Each year, diabetes is the cause of about 36,000 deaths among Americans and is a contributing cause in another 95,000 deaths.
Heart Disease and Strokes
  • People with diabetes are two to four times more likely to have heart disease and two to six times more likely to have a stroke than people who do not have diabetes.
Kidney Disease
  • Ten percent of all people with diabetes develop end-stage kidney disease (where a person requires dialysis or a kidney transplant to live).
  • Nearly 25 percent of all new patients with end-stage renal disease have diabetes.

Source: Department of Health and Human Services, 1988.

Problem: Title worded too generally; set in type that is too small and too faint.

but

Table 3. Effects of Diabetes

Deaths
  • Each year, diabetes is the cause of about 36,000 deaths among Americans and is a contributing cause in another 95,000 deaths.
Heart Disease and Strokes
  • People with diabetes are two to four times more likely to have heart disease and two to six times more likely to have a stroke than people who do not have diabetes.
Kidney Disease
  • Ten percent of all people with diabetes develop end-stage kidney disease (where a person requires dialysis or a kidney transplant to live).

Solution: Editor words title more specifically; uses larger, darker type.

*Terms marked with an asterisk are defined in the Dictionary.

- Nearly 25 percent of all new patients with end-stage renal disease have diabetes.

Source: Department of Health and Human Services, 1988.

SHOW PARTS OF THE WHOLE IN THE STUB COLUMN.

The stub is the first column to the left in a table. It usually labels items described in horizontal rows to the right. Show parts of the whole in the stub column by

- Indentation

or

- Change in size or darkness of type

not

Sponsor
Power Light Consortia
New York Power
Niagara Power
Syracuse Power
Others
U.S. Department of Energy
Edison Research Institute

Problem: Stub doesn't distinguish between categories and subcategories.

but

Sponsor
Power Light Consortia
New York Power
Niagara Power
Syracuse Power
Others
U.S. Department of Energy
Edison Research Institute

Solution: Editor indents "New York Power," "Niagara Power," "Syracuse Power," and "Others" to show they are parts of a larger category, "Power Light Consortia."

USE FIELD COLUMN HEADS TO SHOW PRIMARY COMPARISONS HORIZONTALLY.

Most people overview tables by first scanning them across, then down. Take advantage of this by showing primary comparisons horizontally in the field columns (columns to right of the stub). Distinguish column headings* with boldface, white space, shading, or reversal (white on dark background instead of black on white). Set off supporting column headings with rules or straddle lines.

*Terms marked with an asterisk are defined in the Dictionary.

not

Table 1. Death Rates from Accidents and Violence: 1960 to 1977 (per 100,000 population)

Sex, Cause of Death, and Age	White				Black and Other			
	1960	1970	1975	1977	1960	1970	1975	1977
Male, total	91.3	101.9	96.9	96.8	136.6	174.3	157.3	144.2
Motor vehicle	31.5	39.1	32.2	34.1	34.4	44.3	33.8	33.8
Other accidents	38.6	38.2	35.5	32.7	60.6	60.7	50.2	45.4
Suicide	17.6	18.0	20.1	21.4	7.2	8.5	10.6	11.4
Homicide	3.6	6.8	9.1	8.7	34.5	60.8	62.6	53.6
15–24 years	105.2	130.7	129.7	135.4	147.8	224.3	179.5	157.6
25–44 years	88.5	107.1	103.4	102.5	200.1	275.4	256.4	230.1
45–64 years	116.6	121.4	104.0	98.0	185.1	236.8	203.3	184.8
65 years and over	223.5	216.9	185.3	176.5	202.0	218.0	200.5	188.4

Source: U.S. National Center for Health Statistics.

Problem: Data not grouped vividly in column heads.

Solution: Editor adds straddle rules, boldface, and larger type size to group data in column heads.

but

Table 1. Death Rates from Accidents and Violence: 1960 to 1977 (per 100,000 population)

Sex, Cause of Death, and Age	White				Black and Other			
	1960	1970	1975	1977	1960	1970	1975	1977
Male, total	91.3	101.9	96.9	96.8	136.6	174.3	157.3	144.2
Motor vehicle	31.5	39.1	32.2	34.1	34.4	44.3	33.8	33.8
Other accidents	38.6	38.2	35.5	32.7	60.6	60.7	50.2	45.4
Suicide	17.6	18.0	20.1	21.4	7.2	8.5	10.6	11.4
Homicide	3.6	6.8	9.1	8.7	34.5	60.8	62.6	53.6
15–24 years	105.2	130.7	129.7	135.4	147.8	224.3	179.5	157.6
25–44 years	88.5	107.1	103.4	102.5	200.1	275.4	256.4	230.1
45–64 years	116.6	121.4	104.0	98.0	185.1	236.8	203.3	184.8
65 years and over	223.5	216.9	185.3	176.5	202.0	218.0	200.5	188.4

Source: U.S. National Center for Health Statistics.

ALIGN COLUMNS VERTICALLY BY A COMMON ELEMENT.

In general, align words on left, numbers on the decimal or units digit, and groups on a common sign (plus-minus, multiplication, en dash*).

not

3×10^{10}
37×10^{10}
1.92×10^{10}

but

3×10^{10}
37×10^{10}
1.92×10^{10}

Problem: Column aligned poorly.

Solution: Editor aligns on multiplication sign.

For ranges, align (1) on the dash representing the range (usually an en dash), (2) on the word "to," or (3) on the left.

not

Absorptivity
0.039–0.057
0.018
0.97
0.096–0.29

Problem: Table poorly aligned.

but

Absorptivity
0.039–0.057
0.018
0.97
0.096–0.29

Solution: Editor aligns ranges by en dashes and centers; aligns single numbers on decimal point and centers.

If the column contains both ranges and single numbers of like units, align ranges on the dash and center items; align single numbers on the decimal point or unit digit.

not

Temperature Range, °C
250–600
40–350
1300–3000
100
0

Problem: Poor alignment in column.

but

Temperature Range, °C
250–600
40–350
1300–3000
100
0

Solution: Editor aligns ranges on dash and centers; aligns single numbers on unit digit and centers.

If figures represent different units of measure, or include both worded items and numbers, align on left.

*Terms marked with an asterisk are defined in the Dictionary.

USE ZEROS AND ROUNDED NUMBERS IN TABLES FOR PRECISION, NOT DECORATION.

Do **not** use zeros to the right side of the decimal simply to fill a column entry in a table. Numbers in tables convey concise information. If you record the width of a piece of wood as 2.16″, you mean the measure is closer to 2.16″ than 2.15″ or 2.17″. If you measure the width as 2.00″, you mean by the two zeros that the width is closer to 2.00″ than 1.99″ or 2.01″. As a rule, the uncertainty of the measurement is one unit in the far right digit. When you write a number, an uncertainty is implied by the form in which you write it. The form implies not only a value for the quantity, but an uncertainty about that quantity.

A calculator cannot decide significant digits. For instance, if you weigh a sample at 98.3 g, determine that the amount of oxygen weighs 5.7 g, and ask what percent by weight oxygen is in the sample, you could calculate the fraction $5.7/98.3 \times 100$ to get a percentage. A calculator might give you the number 5.798575, but the number would not be meaningful, as the percentage to the nearest millionth of a percent is not known. How many decimals can be used? A simple rule is that you can retain no more digits than those you started with. In this case, one number had 2 digits (5.7 g), the other 3 (98.3 g). The answer is no more accurate than the least accurate number you started with. Therefore 5.79 becomes 5.8 percent.

USE FOOTNOTES* FOR EXPLANATORY DETAILS, ABBREVIATIONS, SOURCES, STATISTICAL SIGNIFICANCE, AND UNITS OF MEASURE THAT DON'T FIT IN COLUMN HEADS.

Footnotes appear below tables, usually in a smaller typeface. Most academic publications order them with the symbols *, †, ‡, §, ‖, followed by superscript italic letters ([a, b, c]). Many publications simply use the word NOTE.

not

Table 5. Permanent Withdrawal from Study Medications

	Treatment Group			
Reason	PLACEBO (N = 139)	ASA (N = 130)	SULF (N = 140)	ASA + SULF (N = 137)
Patient-initiated	21/6	21/8	26/5	31/5
Side effects	11/6	15/8	12/5	19/5
Compliance	10/0	6/0	14/0	12/0
Physician-initiated	13/2	17/2	14/4	12/4
Medical	7/2	10/2	7/4	9/4
Diagnostic	6/0	7/0	7/0	3/0
Total	34/8	38/10	40/9	43/9

Problem: No explanation of pairings of numbers separated by solidus. No explanation of ASA or SULF.

*Terms marked with an asterisk are defined in the Dictionary.

but

Table 5. Permanent Withdrawal from Study Medications

| Reason | Treatment Group* | | | |
	PLACEBO (N = 139)	ASA (N = 130)	SULF (N = 140)	ASA + SULF (N = 137)
Patient-initiated	21/6	21/8	26/5	31/5
Side effects	11/6	15/8	12/5	19/5
Compliance	10/0	6/0	14/0	12/0
Physician-initiated	13/2	17/2	14/4	12/4
Medical	7/2	10/2	7/4	9/4
Diagnostic	6/0	7/0	7/0	3/0
Total	34/8	38/10	40/9	43/9

*Each pair of numbers separated by solidus indicates first the number of patients in the group who stopped taking both types of pills and then the additional number of patients who stopped taking aspirin or its placebo only. ASA denotes aspirin and SULF sulfinpyrazone.
Source: The New England Journal of Medicine.

Solution: Editor uses footnotes to explain pairings of numbers and abbreviations; adds source note.

not

Table 2. Base-Line Characteristics of the Four Groups of Patients.

| Characteristic | Treatment Group | | | |
	PLACEBO (N = 139)	ASA (N = 139)	SULF (N = 140)	ASA + SULF (N = 137)
Age (mean yr)	57.0	56.7	57.8	56.6
Male	73.4	66.9	74.3	78.1
Angina (past 1–12 mo)	86.3	84.9	87.1	82.5
Prior myocardial infarction	43.2	36.0	42.1	42.3
Prior coronary artery bypass	15.1	5.8	3.6	10.2

Problem: Author has not explained that values in the table represent percentages of patients.

but

Table 2. Base-Line Characteristics of the Four Groups of Patients. *

| Characteristic | Treatment Group | | | |
	PLACEBO (N = 139)	ASA (N = 139)	SULF (N = 140)	ASA + SULF (N = 137)
Age (mean yr)	57.0	56.7	57.8	56.6
Male	73.4	66.9	74.3	78.1
Angina (past 1–12 mo)	86.3	84.9	87.1	82.5
Prior myocardial infarction	43.2	36.0	42.1	42.3
Prior coronary artery bypass	15.1	5.8	3.6	10.2

Solution: Editor uses footnote for explanation; adds source note.

*Values are percentages of patients except where otherwise indicated.
Source: New England Journal of Medicine.

*Terms marked with an asterisk are defined in the Dictionary.

not

Table 1. Characteristics of Male Veterans and Nonveterans Age 18 and Over, November 1987, not Seasonally Adjusted [Percent Distribution]

| | | Veterans | | | | | |
| | | Vietnam-Era Veterans | | | Other War Periods | Other Service Periods | Nonveterans |
Characteristic	Total	Total	Vietnam Theater	Outside Vietnam Theater			
Total (thousands)	25,521	7,902	3,835	4,067	12,612	5,007	57,898
Race or ethnicity:							
White	90.1	88.8	88.8	88.8	91.8	87.6	85.2
Black	8.4	9.3	9.6	9.0	7.0	10.7	11.1
Hispanic	3.1	3.8	4.8	3.0	2.4	3.7	9.3
Age:							
18–24	1.1	—	—	—	—	5.9	21.3
25–34	9.6	11.1	6.4	15.5	—	31.4	31.7
25–29	4.1	.5	.2	.9	—	20.1	16.3
30–34	5.5	10.5	6.2	14.6	—	11.3	15.5
35–44	22.8	66.8	70.6	63.3	—	10.8	18.8
35–39	10.5	30.9	32.2	29.6	—	4.9	11.1
40–44	12.3	35.9	38.4	33.7	—	5.9	7.7
45–54	18.3	15.9	16.3	15.4	9.7	43.7	11.5
55–64	26.7	4.9	5.4	4.4	48.5	6.1	5.9
65 and over	21.5	1.4	1.4	1.4	41.8	2.2	10.8
Disability status:							
Not disabled	87.3	86.0	83.1	88.8	86.1	92.3	—
Disabled, total	9.2	10.3	13.8	6.9	10.7	3.7	—
Less than 30 percent	5.2	5.9	7.9	4.1	5.9	2.3	—
30 to 50 percent	2.1	2.2	2.7	1.6	2.6	1.0	—
60 percent or greater	1.3	1.6	2.3	.9	1.5	.3	—
Presence of disability not reported	3.5	3.7	3.1	4.3	3.2	4.0	—

Problem: Author has not explained dashes in table. Author has not explained why categories of disability ratings sometimes do not sum to totals or why details of racial and Hispanic-origin groups do not sum to totals.

but

Solution: Editor explains what dashes mean and why categories do not sum to totals.

Table 1. Characteristics of Male Veterans and Nonveterans Age 18 and Over, November 1987, not Seasonally Adjusted [Percent Distribution]

| Characteristic | Total | Veterans[1] | | | | | Nonveterans |
| | | Vietnam-Era Veterans | | | Other War Periods | Other Service Periods | |
		Total	Vietnam Theater	Outside Vietnam Theater			
Total (thousands)	25,521	7,902	3,835	4,067	12,612	5,007	57,898
Race or ethnicity:							
White	90.1	88.8	88.8	88.8	91.8	87.6	85.2
Black	8.4	9.3	9.6	9.0	7.0	10.7	11.1
Hispanic	3.1	3.8	4.8	3.0	2.4	3.7	9.3
Age:							
18–24	1.1	—	—	—	—	5.9	21.3
25–34	9.6	11.1	6.4	15.5	—	31.4	31.7
25–29	4.1	.5	.2	.9	—	20.1	16.3
30–34	5.5	10.5	6.2	14.6	—	11.3	15.5
35–44	22.8	66.8	70.6	63.3	—	10.8	18.8
35–39	10.5	30.9	32.2	29.6	—	4.9	11.1
40–44	12.3	35.9	38.4	33.7	—	5.9	7.7
45–54	18.3	15.9	16.3	15.4	9.7	43.7	11.5
55–64	26.7	4.9	5.4	4.4	48.5	6.1	5.9
65 and over	21.5	1.4	1.4	1.4	41.8	2.2	10.8
Disability status:							
Not disabled	87.3	86.0	83.1	88.8	86.1	92.3	—
Disabled, total	9.2	10.3	13.8	6.9	10.7	3.7	—
Less than 30 percent	5.2	5.9	7.9	4.1	5.9	2.3	—
30 to 50 percent	2.1	2.2	2.7	1.6	2.6	1.0	—
60 percent or greater	1.3	1.6	2.3	.9	1.5	.3	—
Presence of disability not reported	3.5	3.7	3.1	4.3	3.2	4.0	—

[1]Because of the aging of the population, there were no longer any Vietnam-era veterans under 25 years of age or any other war veterans under 45 years of age.

[2]Categories of disability ratings may not sum to totals, because specific ratings were not available for some disabled veterans.

NOTE: Details for racial and Hispanic-origin groups will not sum to totals because data for the "other races" group are not presented and Hispanics are included in both the white and black population groups.

Dashes indicate data not available.

KEEP WORDING CONCISE.

If information is identical in a column, try placing the information either in the column head or in a footnote.

not

Table 1. Inorganic Substances: Thermodynamic Data

Substance	Temperature	Atomic or Molecular Weight, amu	Enthalpy of Formation ΔH_f°, kJ/mol	Free Energy of Formation ΔG_f°, kJ/mol
Aluminum				
Al(s)	25°C	26.98	0	0
Al³⁺(aq)	25°C	26.98	−524.7	−481.2
Al₂O₃(s)	25°C	101.95	−1675.7	−1582.3
Al(OH)₃(s)	25°C	78.00	−1276	
AlCl₃(s)	25°C	133.24	−704.2	−628.8

Problem: Column for temperature unnecessary.

but

Table 1. Inorganic Substances: Thermodynamic Data at 25°C

Substance	Atomic or Molecular Weight, amu	Enthalpy of Formation ΔH_f°, kJ/mol	Free Energy of Formation ΔG_f°, kJ/mol
Aluminum			
Al(s)	26.98	0	0
Al³⁺(aq)	26.98	−524.7	−481.2
Al₂O₃(s)	101.95	−1675.7	−1582.3
Al(OH)₃(s)	78.00	−1276	
AlCl₃(s)	133.24	−704.2	−628.8

Solution: Editor places temperature in title.

IN CHARTS AND DISPLAYED LISTS,* USE A DESIGN THAT HELPS READERS SEE AND COMPARE THE BASIC GROUPS WITHIN THE DATA.

not

Problem	Possible Solutions
Copier keeps jamming	See if there are whole sheets or pieces of paper left in the copier. Check paper path. Then check that the AREA 3A handle is in down position, and that AREA 3B handle is pushed in. Perhaps the copy paper is not loaded correctly in the paper trays. Check to make sure the copy paper is in good condition. Try turning the paper over or changing the paper in the trays.

Problem: Poor layout of possible solutions to problem.

*Terms marked with an asterisk are defined in the Dictionary.

but

Problem	Possible Solutions
Copier keeps jamming	**Whole Sheets of Paper Left in Copier?** 1. Check paper path thoroughly. 2. Check that Area 3A handle is **down**. 3. Check that Area 3B handle is pushed **in**. **Copy Paper Loaded Incorrectly in Paper Trays?** 1. Check to make sure copy paper is in good condition. 2. Turn paper over or change paper in trays

Solution: Editor uses boldface column heads to separate problem and solution; boldface questions to state possible solutions; and a displayed, numbered list for solutions.

not

Key features in an exercise cycle

Feature	Function	Comments
Type of resistance control	Changes pressure on wheel to make pedaling harder or easier on cycle.	Be sure you can adjust resistance level while seated
Pressure-wheel type	Tightens a small roller onto cycle wheel to change.	Least expensive. Does not give most accurate resistance levels.
Caliper type	Works like a bicycle brake. Presses pads against wheel rim.	More reliable than pressure wheel. Pads may need replacement occasionally.
Brake-strap type	Tightness of strap around wheel adjusted to change resistance levels.	More reliable than caliper type. Allows adjustments for a great range of resistance levels.
Moving-pad type	Large pads adjust to put pressure against wheel. Used with solid-wheel cycles only.	Most expensive, but gives most reliable resistance levels.
Timer	Shows length of exercise either running time or count-down type.	Check for easy reading. Count-down timer should have easy-to-hear signal.
Pulse sensor	Built into handlebars or worn on body. Some have a read-out device that shows pulse rate directly. Others magnify pulse so you can hear it and count rate.	Can be built into cycle or bought separately. Check for easy reading while cycling.

but

What Are the Key Features in an Exercise Cycle?

Feature	Function	Comments
Resistance control	Changes pressure on wheel to make pedaling harder or easier.	Be sure you can adjust resistance level while seated on cycle.
Pressure wheel	Tightens a small roller onto cycle wheel to change.	Least expensive. Does not give most accurate resistance levels.
Caliper	Works like a bicycle brake. Presses pads against wheel rim.	More reliable than pressure wheel. Pads may need replacement occasionally.
Brake strap	Tightness of strap around wheel adjusted to change resistance levels.	More reliable than caliper type. Allows adjustments for a great range of resistance levels.
Moving pad	Large pads adjust to put pressure against wheel. Used with solid-wheel cycles only.	Most expensive, but gives most reliable resistance levels.
Timer	Shows length of exercise either running time or count-down type.	Check for easy reading. Count-down timer should have easy-to-hear signal.
Pulse sensor	Built into handlebars or worn on body. Some have a read-out device that shows pulse rate directly. Others magnify pulse so you can hear it and count rate.	Can be built into cycle or bought separately. Check for easy reading while cycling.

6/////

EDIT FOR FIGURES THAT IDENTIFY AND EXPLAIN

Figures* are illustrations such as line drawings, graphs, charts, or photographs; captions* or legends are the words that explain the figures. The saying "A picture is worth a thousand words" is regularly proved false by figures that are jumbled, inconsistent, or obscure.

This chapter discusses typical editing strategies for improving the effect of figures and legends.

NUMBER FIGURES; CITE THE NUMBER ON FIRST REFERENCE IN THE TEXT.

Use consecutive arabic numerals. Do not skip numbers or number out of sequence.

not

For cracks around the chimney flue, clean the joint free of all loose caulk and other debris. Using a butyl caulking compound (in either gun form or rope form), seal the joint between the flue and the chimney mortar cap.

Problem: No figure number, source note, or citation.

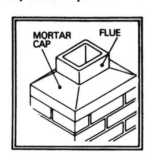

---

*Terms marked with an asterisk are defined in the Dictionary.

but

For cracks around the chimney flue (Fig. 3-1), clean the joint free of all loose caulk and other debris. Using a butyl caulking compound (in either gun form or rope form), seal the joint between the flue and the chimney mortar cap.[8]

Solution: Editor supplies figure number, source note, and figure citation.

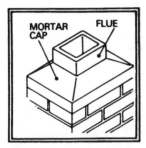

Fig. 3.1. Sealing Cracks. *Source:* U.S. Department of Agriculture.

USE CAPTIONS, LABELS, AND ARROWS TO IDENTIFY AND EXPLAIN FIGURES.

not

Place the new flashing apron (*Fig. 3-2a*) over the old flashing and press into place.

Problem: Figure has number but no explanation.

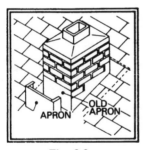

Fig. 3.2a.

but

Place the new flashing apron (*Fig. 3-2b*) over the old flashing and press into place.[9]

Solution: Editor provides caption and callout.*

Fig. 3.2b. Replacing Flashing. *Source:* U.S. Department of Agriculture.

*Terms marked with an asterisk are defined in the Dictionary.

not

Problem: Detailed figure numbered, but not explained.

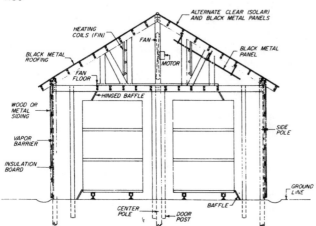

ALTERNATE CLEAR (SOLAR)
AND BLACK METAL PANELS

HEATING
COILS (FIN)

FAN

BLACK METAL
ROOFING

BLACK METAL
PANEL

MOTOR

FAN
FLOOR

HINGED BAFFLE

WOOD OR
METAL
SIDING

SIDE
POLE

VAPOR
BARRIER

INSULATION
BOARD

GROUND
LINE

CENTER
POLE

DOOR
POST

BAFFLE

Fig. 3.3a.

but

Solution: Editor adds informative caption: Fig. 3-3 Cross section of a typical forced-air dryer.

ALTERNATE CLEAR (SOLAR)
AND BLACK METAL PANELS

HEATING
COILS (FIN)

FAN

BLACK METAL
ROOFING

BLACK METAL
PANEL

MOTOR

FAN
FLOOR

HINGED BAFFLE

WOOD OR
METAL
SIDING

SIDE
POLE

VAPOR
BARRIER

INSULATION
BOARD

GROUND
LINE

CENTER
POLE

DOOR
POST

BAFFLE

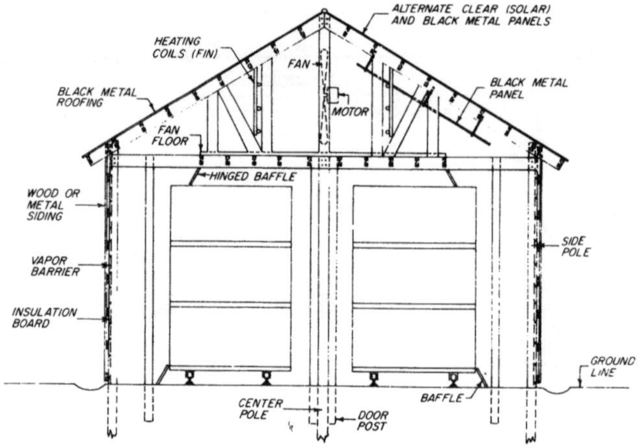

Fig. 3.3b. Cross Section of a Typical Forced-Air Dryer. *Source:* U.S. Department of Agriculture.

EXPLAIN DIVISIONS OF COMPLICATED FIGURES. KEEP WORDING PARALLEL.*

not

Problem: Captions do not explain division in figure.

Fig. 3.4a. Repairing Door.

but

Solution: Editor writes captions that explain division; uses parallel wording.

Fig. 3.4b. Repairing Door. (a) Remove hinges. (b) Plane door. *Source:* U.S. Department of Agriculture.

*Terms marked with an asterisk are defined in the Dictionary.

not

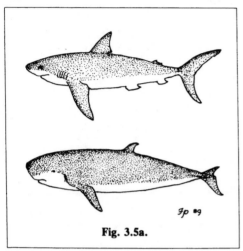

Fig. 3.5a.

but

Fig. 3.5b. The pygmy sperm whale (*bottom*) closely resembles the mackerel shark. *Source:* Flora Pomeroy, *Science Notes*, Spring 1990.

EXPLAIN ALL ABBREVIATIONS,* ACRONYMS, AND SYMBOLS.

Problem: No explanation of abbreviation used in figure; no explanation of figure division; no source note.

not

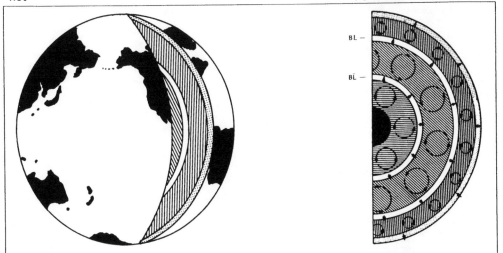

Fig. 3.6a. An orange-segment-like section of the Earth reveals the thin crust, the upper mantle, the pervoskite lower mantle, the liquid outer core, and the solid inner core.

Solution: Editor adds explanation of abbreviation BL; identifies parts of the figure with word "right" and arrows; adds source note.

but

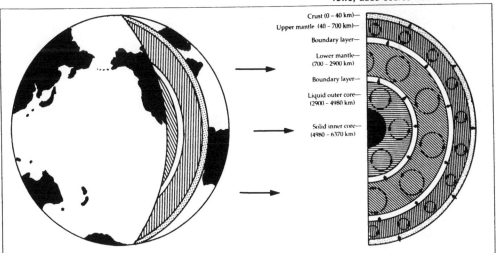

Crust (0 – 40 km)—
Upper mantle (40 – 700 km)—
Boundary layer—
Lower mantle— (700 – 2900 km)
Boundary layer—
Liquid outer core— (2900 – 4980 km)
Solid inner core— (4980 – 6370 km)

Fig. 3.6b. An orange-segment-like section of the Earth (*right*) reveals the thin crust, the upper mantle, the boundary layer (BL), the pervoskite lower mantle, the liquid outer core, and the solid inner core. *Source:* Amy Sibiga, *Science Notes,* Fall 1989.

*Terms marked with an asterisk are defined in the Dictionary.

not   *but*

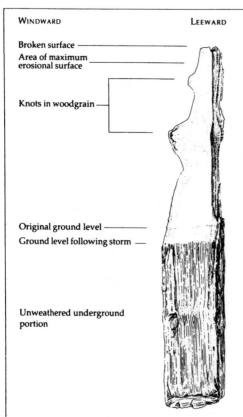

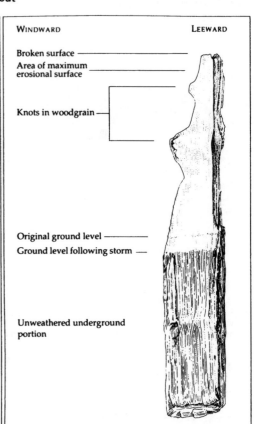

WINDWARD   LEEWARD

Broken surface ─────────────

Area of maximum
erosional surface ─────────────

Knots in woodgrain ─

Original ground level ─────────

Ground level following storm ─

Unweathered underground
portion

Fig. 3.7a. Cedar fencepost sculpted by sand and dirt.

WINDWARD   LEEWARD

Broken surface ─────────────

Area of maximum
erosional surface ─────────────

Knots in woodgrain ─

Original ground level ─────────

Ground level following storm ─

Unweathered underground
portion

Fig. 3.7b. Cedar fencepost sculpted by sand and dirt. The leeward side (right) and the buried lower half were protected from wind-blown grit. The windward side (upper left) was cut more or less, depending on such factors as the hardness of the wood and the force with which the particles hit. *Source:* Ellen Bennett, *Science Notes*, Spring 1989.

EXPLAIN ANY ARROWS, POINTERS, OR SPECIAL SIGNS.

not

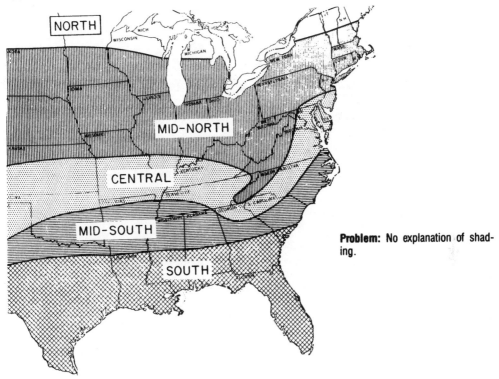

Problem: No explanation of shading.

Fig. 3.8a. Air drying map for the Eastern United States.

but

Solution: Editor adds legend.

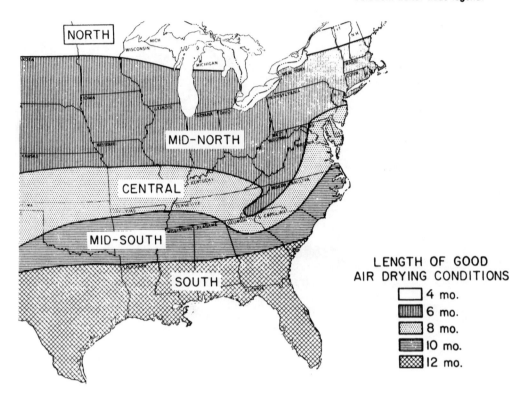

Fig. 3.8b. Air drying map for the Eastern United States. *Source:* U.S. Department of Agriculture.

not

Problem: Dotted line unexplained.

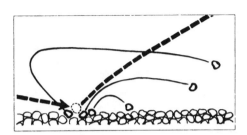

Fig. 3.9a. A sand grain hits a bed of sand, rebounds, and simultaneously knocks other grains from the bed. Some are caught and carried along by the wind.

but

Solution: Editor explains significance of dotted line.

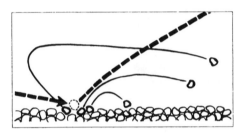

Fig. 3.9b. A sand grain (dotted line) hits a bed of sand, rebounds, and simultaneously knocks other grains from the bed. Some are caught and carried along by the wind. *Source:* Judy Ward, *Science Notes,* Spring 1989.

MAKE SURE THE INFORMATION IN THE CAPTION AND CALLOUT IS CONSISTENT WITH THE INFORMATION IN THE TEXT.

not

The mousetrap has five parts:

1) The *platform*, a rectangular wooden base.
2) The *clamp*, a U-bar mounted to the platform.
3) The *spring*, a metal coil attached to the clamp and platform.
4) The *bar*, a metal rod used to restrain the clamp.
5) The *bait plate*, which holds the bait and rod.

Problem: Callout inconsistent with text ("bait plate" in text, "trigger" in callout).

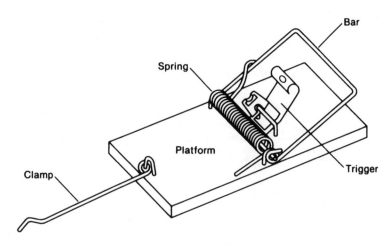

Fig. 3.10a. Parts of the Mousetrap.

but

Solution: Editor reconciles callout and text.

The mousetrap has five parts:

1) The *platform*, a rectangular wooden base.
2) The *clamp*, a U-bar mounted to the platform.
3) The *spring*, a metal coil attached to the clamp and platform.
4) The *bar*, a metal rod used to restrain the clamp.
5) The *bait plate*, which holds the bait and rod.

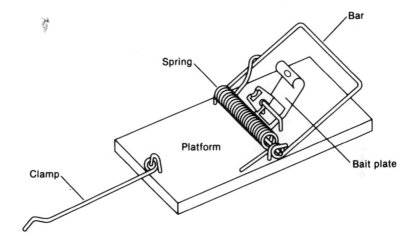

Fig. 3.10b. Parts of the Mousetrap.

7/////

EDIT FOR COMPLETENESS, ACCURACY, AND CONSISTENCY

Errors that leave copy incomplete, inaccurate, or inconsistent creep into final drafts of documents. For instance, in last-minute changes the author may add or delete data without making the corresponding changes to the figures. In rewording text and adjusting tables during successive drafts, the author may end up discussing numbers in the narrative that do not match numbers in the displays. Definitions of abbreviations vanish; numbers reverse themselves. A term appears as two words on page 7, but as one word on page 11. Pages are misnumbered, or in the wrong order. The table of contents is not consistent with headings and subheadings in the text.

This chapter discusses some of the more typical errors editors look for when they check content for completeness, accuracy, and consistency.

CHECK FOR CONSISTENCY BETWEEN TABLE OR FIGURE DESIGNATOR (TABLE 1) AND IN-TEXT CITATION (SEE TABLE 1).

not

Divorce rates have shown a long-term increase in most industrial nations since around the turn of the century. (See Table 1.) After accelerating during the 1970s, the rates reached in the 1980s are probably the highest in the modern history of these nations. . . .

Problem: Table citation (See Table 1) does not match table designator (Table 3).

but

Divorce rates have shown a long-term increase in most industrial nations since around the turn of the century. (See Table 3.) After accelerating during the 1970s, the rates reached in the 1980s are probably the highest in the modern history of these nations. . . .[10]

Solution: Editor checks table citation against table designator and corrects.

Table 3. Marriage and Divorce Rates in 10 Countries,
Selected Years, 1960–86

Country	1960	1970	1980	1986
	Marriage rates (per 1,000 population, ages 15 to 64)			
United States	14.1	17.0	[1]15.9	15.1
Canada	12.4	14.3	11.8	10.2
Japan	14.5	14.4	9.8	8.6
Denmark	12.2	11.5	7.9	9.0
France	11.3	12.4	9.7	7.3
Germany	13.9	11.5	8.9	8.7
Italy	11.7	11.3	8.7	7.5
Netherlands	12.7	15.2	9.6	8.7
Sweden	10.2	8.2	7.1	7.2
United Kingdom	11.5	13.5	11.6	10.6
	Divorce rates (per 1,000 married women)			
United States	9.2	14.9	22.6	21.2
Canada	1.8	6.3	10.9	12.9
Japan	3.6	3.9	4.8	5.4
Denmark	5.9	7.6	11.2	12.8
France	2.9	3.3	6.3	8.5
Germany	3.6	5.1	6.1	8.3
Italy	([2])	1.3	.8	1.1
Netherlands	2.2	3.3	7.5	8.7
Sweden	5.0	6.8	11.4	11.7
United Kingdom	2.0	4.7	12.0	12.9

[1]Beginning in 1980, includes unlicensed marriages registered in California.
[2]Not available.
Sources: Statistical Office of the European Communities, *Demographic Statistics, 1988;* and various national sources.

Table 3. Marriage and Divorce Rates in 10 Countries,
Selected Years, 1960–86

Country	1960	1970	1980	1986
	Marriage rates (per 1,000 population, ages 15 to 64)			
United States	14.1	17.0	¹15.9	15.1
Canada	12.4	14.3	11.8	10.2
Japan	14.5	14.4	9.8	8.6
Denmark	12.2	11.5	7.9	9.0
France	11.3	12.4	9.7	7.3
Germany	13.9	11.5	8.9	8.7
Italy	11.7	11.3	8.7	7.5
Netherlands	12.7	15.2	9.6	8.7
Sweden	10.2	8.2	7.1	7.2
United Kingdom	11.5	13.5	11.6	10.6
	Divorce rates (per 1,000 married women)			
United States	9.2	14.9	22.6	21.2
Canada	1.8	6.3	10.9	12.9
Japan	3.6	3.9	4.8	5.4
Denmark	5.9	7.6	11.2	12.8
France	2.9	3.3	6.3	8.5
Germany	3.6	5.1	6.1	8.3
Italy	(²)	1.3	.8	1.1
Netherlands	2.2	3.3	7.5	8.7
Sweden	5.0	6.8	11.4	11.7
United Kingdom	2.0	4.7	12.0	12.9

¹Beginning in 1980, includes unlicensed marriages registered in California.
²Not available.
Sources: Statistical Office of the European Communities, *Demographic Statistics, 1988;* and various national sources.

CHECK FOR CONSISTENCY BETWEEN DATA DISCUSSED IN NARRATIVE AND DATA DISPLAYED IN TABLES.

not

According to Table 3, a trend toward fewer marriages is plain in all of the countries studied, although the timing of this decline differs from country to country. In Scandinavia and Germany, for example, the downward trend in the marriage rate was already evident in the 1960s; and in the United States, Canada, Japan, Holland, and the United Kingdom. . .

Divorce rates have shown a long-term increase in most industrial nations since around the turn of the century. After accelerating during the 1970s, the rates reached in the 1980s are probably the highest in the modern history of these nations. While a very large proportion of Americans marry, their marital breakup rate is by far the highest among the developed countries. Based on recent divorce rates, the chances of a first American marriage ending in divorce are today about one in two; the corresponding ratio in Europe is about one in three to one in four.

Liberalization of divorce laws came to the United States well

Problem: Table says "Netherlands," not "Holland."

before it occurred in Europe, but such laws were loosened in most European countries beginning in the 1970s, with further liberalization taking place in the 1980s. Consequently, divorce rates are rising rapidly in many European countries. By 1986, the rate had quadrupled in the Netherlands and tripled in France. The sharpest increase occurred in the United Kingdom, where the marital breakup rate increased sixfold. Although divorce rates continued to rise in Europe in the 1980s, the increase in the United States abated, and the rate in 1986 was slightly below that recorded in 1980. In Canada, although divorce rates remain considerably lower than in the United States, the magnitude of the increase since 1960 has been greater than in the United Kingdom.

Italy is the only European country studied in which the divorce rate remains low, and divorce laws have not been liberalized there. Japan's divorce rates are lower than in all other countries except Italy. . . .

Divorce rates understate the extent of family breakup in all countries: marital separations are not covered by the divorce statistics, and these statistics also do not capture the breakup of families in which the couple is not legally married. Studies show that in Sweden, the breakup rate of couples in consensual unions is three times the dissolution rate of married couples.

but

According to Table 3, a trend toward fewer marriages is plain in all of the countries studied, although the timing of this decline differs from country to country. In Scandinavia and Germany, for example, the downward trend in the marriage rate was already evident in the 1960s, and in the United States, Canada, Japan, the Netherlands, and the United Kingdom. . . .

Divorce rates have shown a long-term increase in most industrial nations since around the turn of the century. After accelerating during the 1970s, the rates reached in the 1980s are probably the highest in the modern history of these nations. While a very large proportion of Americans marry, their marital breakup rate is by far the highest among the developed countries. Based on recent divorce rates, the chances of a first American marriage ending in divorce are today about one in two; the corresponding ratio in Europe is about one in three to one in four.

Liberalization of divorce laws came to the United States well before it occurred in Europe, but such laws were loosened in most European countries beginning in the 1970s, with further liberalization taking place in the 1980s. Consequently, divorce rates are rising rapidly in many European countries. By 1986, the rate had quadrupled in the Netherlands and almost tripled in France over the levels recorded in 1960. The sharpest increase occurred in the United Kingdom, where the marital breakup rate increased sixfold. Although divorce rates continued to rise in Europe in the 1980s, the increase in the United States abated, and the rate in 1986 was slightly below that recorded in

Problem: According to the table, rate has not quite tripled in France. Also, statement, "By 1986, the rate had quadrupled in the Netherlands and tripled in France," does not answer the question, "In relation to what other rates?"

Solution: Editor changes "Holland" to "Netherlands."

Editor queries author; qualifies* "quadrupled in the Netherlands and tripled in France" to "quadrupled in the Netherlands and almost tripled in France over the levels recorded in 1960."

1980. In Canada, although divorce rates remain considerably lower than in the United States, the magnitude of the increase since 1960 has been greater than in the United Kingdom.

Italy is the only European country studied in which the divorce rate remains low, and divorce laws have not been liberalized there. Japan's divorce rates are lower than in all other countries except Italy. . . .

Divorce rates understate the extent of family breakup in all countries: marital separations are not covered by the divorce statistics, and these statistics also do not capture the breakup of families in which the couple is not legally married. Studies show that in Sweden, the breakup rate of couples in consensual unions is three times the dissolution rate of married couples.[11]

CHECK FOR CONSISTENCY BETWEEN FIGURE CAPTIONS AND NARRATIVE.

not

Room air-conditioners perform better and last longer with simple cleaning and maintenance twice a year: before the cooling season begins and midway through the summer.

First, unplug the air-conditioner. Vacuum the front grille and any accessible components; then dust the surfaces with a soft cloth or wipe them clean with a cloth and household spray cleaner.

Next, remove the grille to check the air filter, which is usually just inside. (See Fig. 7-1a) If you are not sure how to remove the grille, read the owner's manual. Some grilles are held in place by screws; others are made of flexible plastic and pop away from the unit when pressed along the top or pried carefully with a putty knife. . . .

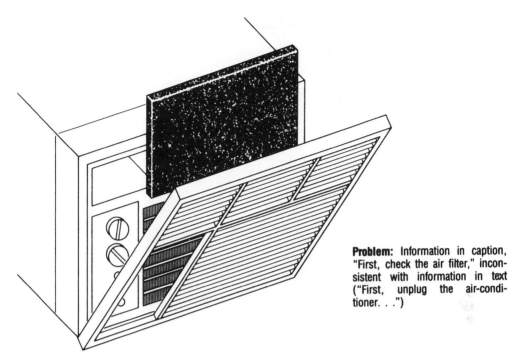

Problem: Information in caption, "First, check the air filter," inconsistent with information in text ("First, unplug the air-conditioner. . .")

Fig. 7.1a. Cleaning and maintenance will improve performance of an air-conditioner. First, check the air filter.

but

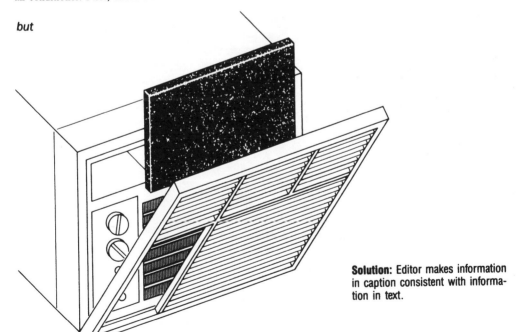

Solution: Editor makes information in caption consistent with information in text.

Fig. 7.1b. First, unplug air-conditioner, vacuum the front grille, and wipe surfaces clean. Then remove the grille to check the air filter, which is usually just inside.

CHECK ACCURACY OF PEOPLE'S NAMES, TITLES, AFFILIATIONS, AND ANY ABBREVIATIONS OF THIS INFORMATION.

Except for initials, given names are rarely abbreviated in text or bylines. Titles, too, are rarely abbreviated except when they precede the full name.

not

Mr. Smith, speaking on behalf of the Veterans' Administration, urged that funding for the project be increased.

but

Mr. Smith, speaking on behalf of the Veterans Administration, urged that funding for the project be increased.

Problem: "Veterans Administration" does not use the apostrophe in its name.

Solution: Editor checks preferred usage for "Veterans Administration"; deletes apostrophe.

not

Asst. Prof. Kennedy, speaking on behalf of the university. . . .

but

Assistant Professor Kennedy, speaking on behalf of the university. . . .

Problem: Titles are not abbreviated before surname alone.

Solution: Editor writes out title.

CHECK ACCURACY OF NAMES OF ORGANIZATIONS* OR AGENCIES.

not

The process was first introduced by Smith Kline & French Laboratories.
He was a speaker for The Centers for Disease Control.

but

The process was first introduced by SmithKline Beecham.
He was a speaker for the Centers for Disease Control.

Problems: "Smith Kline & French Laboratories" not the current name of company. "Centers for Disease Control" does not use "The" as part of its title.

Solution: Editor checks name of company/agency and revises.

CHECK ACCURACY OF PUBLICATION NAMES.

not

Publications include *The Monthly Labor Review* and *The Economist*.

but

Publications include the *Monthly Labor Review* and *The Economist*.

Problem: Incorrect name of publication.

Solution: Editor checks standard list of publications, finds that *Monthly Labor Review* does not use "The" in its title, and deletes "The."

*Terms marked with an asterisk are defined in the Dictionary.

CHECK ALL ABBREVIATIONS* FOR TECHNICAL TERMS.

not

MMIP

but

MMPI

Problem: Author reversed two letters in abbreviation.

Solution: Editor corrected abbreviation for Minnesota Multiphasic Personality Inventory.

CHECK ALL TECHNICAL NOMENCLATURE.*

not

They assayed the octyldodecanol

but

They assayed the octyl dodecanol.

Problem: Preferred form of "octyl dodecanol" is as two words.

Solution: Editor checks USP; gives preferred form of term.

CHECK TO MAKE SURE REFERENCES ARE COMPLETE AND CONSISTENT.

not

[1]David Popenoe, *Disturbing the Nest: Family Change and Decline in Modern Societies* (New York, Aldine DeGruyter, 1988), p. 283.
[2]Jean-Paul Sardon, "Evolution de la nuptialite et de la divortialite en Europe depuis la fin des annees 1960" [Movement in Marriage and Divorce Rates in Europe since the Late 1960's], *Population*, no. 3, May-June 1986, pp. 463–482. *Population* is the journal of the French National Institute of Demographic Studies.
[3]Popeno, *Disturbing the Next*, p. 173.

Problem: Citation 1 lists author's name as "Popenoe"; citation 3 lists author as "Popeno." Citation 1 lists book as *Disturbing the Nest*; citation 3 lists book as *Disturbing the Next*. Citation 1 indented; citations 2 and 3 not indented.

but

[1]David Popenoe, *Disturbing the Nest: Family Change and Decline in Modern Societies* (New York, Aldine DeGruyter, 1988), p. 283.
[2]Jean-Paul Sardon, "Evolution de la nuptialite et de la divortialite en Europe depuis la fin des annees 1960" [Movement in Marriage and Divorce Rates in Europe since the Late 1960's], *Population*, no. 3, May-June 1986, pp. 463–482. *Population* is the journal of the French National Institute of Demographic Studies.
[3]Popenoe, *Disturbing the Nest*, p. 173.

Solution: Editor indents all citations same amount. Verifies spelling of author's name and title and corrects.

---

*Terms marked with an asterisk are defined in the Dictionary.

**CHECK FOR CONSISTENCY IN SIZE AND FACE OF TYPE
USED IN HEADINGS AND SUBHEADINGS.**

not

Demographic Background

Major demographic and sociological changes directly influencing family composition have taken place in this century, with
the pace of change accelerating in the past two decades. Almost all developed countries have seen changes of four principal types: A decline in fertility rates, the aging of the population, an erosion of the institution of marriage, and a rapid increase in childbirths out of wedlock. Each of these four trends
has played a part in the transformation of the modern family.

Fertility rates Over the past century, women in industrialized
countries have moved to having fewer children—that is, to
lower fertility rates. The decline was, in many cases, interrupted by the post-World War II baby boom, but it resumed in
the 1960's. Japan is an exception, in that fertility rates have
declined sharply and almost continuously since the late 1940's,
with no postwar upturn apart from a small recovery and stabilization from the mid-1960's to the early 1970's.

The change in total fertility rates in 10 countries is shown in
Table 1. With the exception of some baby "boomlets" in the
late 1970's and 1980's, total fertility rates in most developed
countries have declined to below the level needed to replace
population deaths, namely, 2.1 children per woman. This
means that the current population will not even replace itself if
current levels of fertility continue. By 1988, fertility rates in
the developed countries fell into a narrow range of from 1.3 to
1.4 children per woman in Germany and Italy to around 1.9 to
2.0 in the United States and Sweden.

Aging of the Population. It is important to consider the age
structure of the population because different arrays of persons
by age result in different household structures across countries.
Mortality, as well as fertility, has declined in the 20th century.
The decline in mortality has been more or less continuous, and
the average age at death has risen considerably in all developed
countries. The decrease in fertility has resulted in a decline in
the proportion of children in the population. However, because
it affected all age groups, the drop in mortality did not have a
major effect on the age structure of populations. In fact, mortality decreased more at younger than at old ages, thereby
offsetting rather than exacerbating the effect of the fertility
decline. Thus, the progressive aging of the population in the
developed countries is attributable primarily to the declining
fertility rates.

Table 2 shows the distribution of the population by age in 10
countries from 1950 to 1990. The proportion of the population

Problem: Writer has not decided on
a consistent format* for subheadings.

The first subhead, *Fertility rates,* is
flush left,* set in italics,* and without end punctuation. Only the first
word uses an initial capital.

The second subheading is flush left,
is set in boldface, and has end
punctuation. Initial capitals are used
for all major words.

*Terms marked with an asterisk are defined in the Dictionary.

in the youngest age group (0–14 years) is declining everywhere, while the proportion of the elderly (age 65 and over) is increasing. Compared with most European countries and Japan, the U.S. and Canadian populations are more youthful, reflecting higher comparative fertility rates. However, in both North American countries the declining fertility rates have produced a sharp drop since 1960 in the share of the population held by the under-age-15 group. With the exception of France, all the European countries and Japan now have less than one-fifth of their total population under 15, with Germany having the lowest proportion.

At the other end of the spectrum, European countries tend to have larger proportions of elderly persons than do the two North American nations. Sweden, Germany, and Denmark all have about the same proportion of elderly as they have children under 15. In contrast, the proportion of children in the United States and Canada is nearly twice as great as the proportion of elderly.

Marriage and divorce. Almost everyone in the United States gets married at some time in his or her life. The United States has long had one of the highest marriage rates in the world, and, even in recent years it has maintained a relatively high rate. For the cohort born in 1945, for example, 95 percent of the men have married, compared with 75 percent in Sweden. The other countries studied ranked somewhere between these two extremes. . . .

The high U.S. marriage rate is, in part, related to the fact that the United States has maintained a fairly low level of nonmarital cohabitation. In Europe—particularly in Scandinavia, but also in France, the United Kingdom, and the Netherlands—there have been large increases in the incidence of unmarried couples living together. This situation is reflected in the lower marriage rates of these countries. . . .

but

Major demographic and sociological changes directly influencing family composition have taken place in this century. . . .

Fertility rates. Over the past century, women in industrialized countries have moved to having fewer children. . . .

Aging of the population. It is important to consider the age structure of the population because different arrays of persons by age result in different household structures across countries. Mortality, as well as fertility, has declined in the 20th century. . . .

Marriage and divorce. Almost everyone in the United States gets married at some time in his or her life. The United States has long had one of the highest marriage rates in the world, and, even in recent years it has maintained a relatively high rate. . .[12]

> Third subheading begins three spaces in, is set in italics, uses initial capital for first word only, and has end punctuation.

> **Solution:** Editor decides on line position, type, punctuation, and capitalization for all subheadings. (In this case, editor chooses italics* for the type, flush left* for the line position, initial capitals only, and periods for end punctuation.) Editor changes subheadings so they conform to guidelines.

*Terms marked with an asterisk are defined in the Dictionary.

CHECK FOR CONSISTENCY IN FORMAT OF DISPLAYED LISTS.*

not

The new pact calls for distribution of the entire money pool without an offset. Other terms include:
• A "me-too" clause, which provides that any more favorable terms negotiated by the Seattle-based engineers be extended to the Wichita engineers.
• Improved pension benefits for active employees equal to the greater of a $30 (was $24) monthly pension rate per year of service, or average earnings in the highest 60 months of the last 120.
• Improved medical benefits for both active employees and retirees, including new coverage for routine physicals and well-baby care and extended coverage for vision care, hospice care, alcohol and drug abuse treatment, and eating disorders.

Lump-sum payments of $500 each will be distributed to employees at the top of the wage progression scale in February 1991 and December 1992, with proportionally smaller payments to employees advancing up the wage progression scale. Other terms include:

   • A $73.55 increase in the employers' $193.95 monthly payment to the health and welfare fund for each worker beginning in June 1991, subject to two additional $20 increases if needed to maintain benefit levels.

   • Changing Easter Sunday and Memorial Day from holidays to floating personal days, effective in 1990.

   • A temporary 3-year cut in Sunday premium pay from time and one-half to time and one-quarter, and in daily overtime pay from time and one-half to straight time.

but

The new pact calls for distribution of the entire money pool without an offset. Other terms include:

   • A "me-too" clause, which provides that any more favorable terms negotiated by the Seattle-based engineers be extended to the Wichita engineers.

   • Improved pension benefits for active employees equal to the greater of a $30 (was $24) monthly pension rate per year of service, or average earnings in the highest 60 months of the last 120.

Problem: Lists are displayed inconsistently.

First displayed list follows text without empty carriage return (white space); bullets* are small (6 pt.) and flush left.*

Second displayed list separated from narrative by carriage return. Bullets larger (8 pt.). Bullets placed three spaces into the line.

Solution: Editor decides on format for displayed lists and revises lists so that they are consistent with format chosen. (In this case, both lists are separated from running text by carriage return. Bullets begin three spaces in from margin. Bullets are 7 pt.)

*Terms marked with an asterisk are defined in the Dictionary.

• Improved medical benefits for both active employees and retirees, including new coverage for routine physicals and well-baby care and extended coverage for vision care, hospice care, alcohol and drug abuse treatment, and eating disorders.

Lump-sum payments of $500 each will be distributed to employees at the top of the wage progression scale in February 1991 and December 1992, with proportionally smaller payments to employees advancing up the wage progression scale. Other terms include:

• A $73.55 increase in the employers' $193.95 monthly payment to the health and welfare fund for each worker beginning in June 1991, subject to two additional $20 increases if needed to maintain benefit levels.

• Changing Easter Sunday and Memorial Day from holidays to floating personal days, effective in 1990.

• A temporary 3-year cut in Sunday premium pay from time and one-half to time and one-quarter, and in daily overtime pay from time and one-half to straight time.

DICTIONARY

A

A, an. Use the indefinite article *a* before a consonant sound: *a level, a unit of measurement, a history* (aspirated "h"). Use the indefinite article *an* before a vowel sound: *an NMR spectrometer, an online system, an 80-year-old building.* Use an indefinite article before each coordinate noun in a series: *Nominated were an electronics engineer, an electrical engineer, and a computer scientist.*

Abbreviations and acronyms. Abbreviations are pronounced as letters (*IBM*), acronyms (*DEW line*) as words. On first reference, use full spelling of the term you want to abbreviate, followed by the shortened form in parentheses. For subsequent references, use the shortened form. *Smokeless tobacco (SLT) poses serious health hazards. Teen-aged males are particularly heavy users of SLT.* Use full wording of all but the most common abbreviations on first use in abstracts and figure captions. Avoid abbreviations in titles. Relax the practice of spelling out abbreviations only when space is very tight (column heads on tables, axis labels on figures), or when the audience is assumed to know the shortened form. For example, among chemists no definitions are needed for RNA, DNA, IR, wt, w/w.

Most abbreviations for units of measure do not use points (*amp, Hz, cal, cm, hr, hp, sec, rad,* but *at. wt, gal., in., no.,* as they may easily be misread as the words "at," "gal," "in," and "no"). If an abbreviation using a

point falls at the end of the sentence, do not add a period.

Singular and plural forms for units of measure are identical; do not use an *s* to indicate a plural: *3 mL* not 3 *mLs.* Form the plural of capitalized abbreviations by adding a lowercase "s"; no apostrophe is necessary: *ICBMs.* Abbreviate all units of measure that follow a number: 31 g/L, 31 g L^{-1} but *The results, measured in milligrams per kilogram, were . . . .*

Do not abbreviate months without dates, states without cities, or days of the week; do not abbreviate the words "day," "week," "month," or "year" in text. Do not abbreviate "United States" unless used as an adjective: *in the United States* but *U.S.-related projects.*

Avoid beginning sentences with abbreviations, but if you must, and if the term begins with a lowercase letter, do not capitalize: *dAMP occurs . . . .*

Above, below. Do not use "the above figure" or "the table below" in manuscripts, as the author rarely has any control over the location of figures or tables in the published version. Use "In Figure 2" or "See Table 2."

Abstracts, informative vs. indicative. An **informative abstract** is a synopsis of the findings and their significance. Include what the author found out and why this information matters. Cast it so that someone within the field will know whether to read the paper or report, and someone outside the field need not read the full paper or report. Write out abbreviations and acronyms on first use; use no references, tables, or figures unless permitted by the professional society. Include information from these categories: objective; scope and limits; background (optional, depending on audience background, and usually compressed to a phrase or sentence); procedure (highly compressed unless the method itself is of interest); results, conclusions, and implications. A maximum of 150–160 words is common. Be specific. (Not *Results are given and conclusions drawn* but *Thirteen of 27 subjects accepted the pseudomemories. Six insisted the events occurred, even when told that the incidents had been fabricated. These findings cast doubt on the use of hypnosis in forensic investigation.*) Do not belabor procedural details; compress them to leave room for the findings.

An **indicative abstract** lists the purpose, scope, and major topics in the report or paper, as a table of contents lists major subjects. It makes no attempt to inform the reader of the findings of each study. An indicative abstract is appropriate when the material to be summarized is disparate or voluminous—for instance, an extensive literature review on microbial metabolism of chlorinated aromatic compounds.

Acronyms. See **Abbreviations and acronyms.**

Active vs. passive voice. The active voice emphasizes the subject of the sentence: *Jones measured the refractive index of the liquid.* The passive voice emphasizes the object: *The refractive index of the liquid was measured.* The passive is formed by a combination of the verb "to be" and the past participle (*was measured*); it may contain a "by" clause at the discretion of the writer (*"by Jones"*).

Use the passive when the performer is irrelevant to, or less important than, the object. *The diskette is composed of an easily magnetized, highly flexible material. The 55-year-old man was isolated from other patients for three days.* Avoid passives when the focus belongs on the subject. Not *It has been recommended by the Board that the measures be adopted* but *The Board recommends that the measures be adopted.* Not *An assessment of the damage was made by the engineering staff* but *The engineering staff assessed the damage.* The active voice offers the advantage of shorter, more direct sentences.

For instructions, use the active voice. Not

Disks should be stored upright but *Store disks upright.* Not *The exposed area of the disk should not be touched* but *Do not touch exposed area of the disk.* Use the passive for a procedure (a document explaining how steps are done). *All laboratory notebooks should be signed, dated, and witnessed at the end of the day.*

Do not shift from active to passive within the same sequence. Not *Depress lever and then the screen is raised* but *Depress lever and raise screen.*

Watch for misplaced or dangling modifiers with a passive. Not *When depressed, the screen may be raised* but *When the lever is depressed, the screen may be raised.*

Avoid these expendable passives: *It has been observed that, it has been concluded that.* Instead, give observations and conclusions. See also **Unnecessary words.**

Adapt vs. adopt. Adapt means "to adjust or modify": *We adapted the IBM software so that it would run on our Macintosh.* Adopt means "to take up, accept, or choose." *The committee adopted a new textbook for first-year chemistry.*

Adverbs, placement of. Try to place adverbs between elements in a compound verb. *We had regularly observed* but not *We hoped to clearly see.* See **Infinitives.**

Affect, effect. "Effect" is the noun. As a noun, it is always marked by "a," "an," or "the." *The light source produced a marked effect on growth.* "Affect" is a verb, meaning "to act upon or influence" (*The light beam did not affect growth*) or "to pretend to" (*The seminar speaker affected an interest in lasers*). "Effect" may also be a verb, meaning "carried out" or "accomplished" in such bureaucratic expressions as *We effected a change in the policy.*

A.M., P.M. (also widely a.m., p.m.). Use numbers with A.M. and P.M.: *4:20 A.M.* (but not *4:20 A.M. in the morning, 2:20 P.M. in the afternoon*). See **Unnecessary words.**

Amount vs. number. Use number, not amount, for countable units. *The number of PCs increases each year* not *The amount of PCs increases each year.* Note "The number is" but "A number are."

Analogy. A form of inference in which it is reasoned that if two things agree with one another in one or more respects, they may agree in other respects. Richard Feynman writes:

When I say "light" in these lectures, I don't mean simply the light we can see, from red to blue. It turns out that visible light is just a part of a long scale that's analogous to a musical scale in which there are notes higher than you can hear and other notes lower than you can hear. The scale of light can be described by numbers—called the frequency. As the numbers get higher, the light goes from red to blue to violet to ultraviolet. . . .[13]

And/or. Avoid.

Apostrophe. Use the apostrophe to show possession (*Darwin's notebooks*), to indicate a contraction (*hadn't*), and to form plurals of letters, symbols, and words referred to as words (*Repetition of e's and i's was common*). In general, no apostrophe is needed for all-capitalized abbreviations or for numbers (*ICBMs, a patient in his 20s, during the 1960s*). Use an apostrophe and "s" to form the possessive singular, including most nouns ending in "s" (*DeVries's paper, Marks's report* but note historical exceptions: *Gauss' law, Bayes' theorem, Stokes' law*, and some double sibilants such as *Jesus'* and *Moses'*).

Use an apostrophe to form the possessive of plural nouns endings in "s" or "es": *The authors' results.* For all other plural nouns, add "s": *men's.* For more than one owner, add the apostrophe to the last name: *Pauling and Wilson's book on quantum mechanics.* Pronominal possessives (*his, hers, its, ours, theirs, yours*) use no apostrophe; indefinite pronouns (*anyone's, someone else's*) do use the apostrophe. Never use an apostrophe to indicate the plural of a name. Note use of possessive in *a month's delay.* See also **Plurals.**

Appendixes. The anglicized form is an increasingly popular replacement for *appendices.* See **Greek and Latin words and expressions.**

Articles, indefinite. See **A, an.**

Articles, scientific. See **Scientific papers.**

Artwork. The figures (photographs, line drawings, graphs, charts) and tables that accompany and illustrate a document. See **Captions, Figures,** and **Tables.**

As much as. Do not use *It is as much, if not more, than we expected.* (See **Parallel con-** struction.) Prefer *It is as much as we expected, if not more.* Also acceptable: *It is as much as, if not more than, we expected.*

Assure. See **Insure vs. ensure, assure.**

Author. Not a verb. *We wrote* not *We authored.*

Averse/adverse. Averse means unwilling. *He was averse to changing the dosage.* Adverse means unfavorable, as in *adverse effects of the medication.*

B

Because of, owing to, due to. *Because of* and *owing to* are used properly as adverbs. Not *Due to its poor sales performance, we stopped producing the model* but *Because of its poor sales performance, we stopped producing the model.* (*Because of* modifies the verb *stopped producing.*) *Due to* is used as an adjective to modify nouns and pronouns: *His promotion was due to hard work.* (*Due to* modifies *promotion.*)

Do not use *due to* because you fear starting a sentence with *because. Because* is a subordinate conjunction, and may be properly used to begin a sentence, provided you follow the "because" clause with an independent clause.

Before vs. prior to. Prefer the shorter "before." Not *We mixed Batch 1 with Batch 2 prior to testing* but *We mixed Batch 1 with Batch 2 before testing.* See **Unnecessary words.**

Between . . . or. Incorrect. Use *between . . . and. The choice is between 5½" disks and 3½" disks.*

Between vs. among. Use "between" for two items. *The consumer has a choice between two models.* Use "among" for more than two items. *The consumer has a choice among hundreds of models.* "Between" may also be used for more than two items when "among" is too vague to draw attention to each party in the exchange. *The talks between IBM, Sony, and Control Data ended in wide disagreement among the participants.*

Between you and I. Incorrect. "I," a subject pronoun, may not be used as an object. The correct expression is "between you and me" ("me" is the object of the preposition "between").

Bi- vs. semi-. "Bi" means two; "semi," half. This distinction is widely confounded in daily use, where "biweekly" (literally "every two weeks") is incorrectly used to mean "twice a week." Given this confusion, prefer "every two weeks" to "biweekly" and "twice a week" to "semiweekly."

Similar confusion exists with "biannual" (literally "every two years"), which many people use to mean "twice a year." Prefer "every two years" to "biannually," and "every six months" to "semiannually" or "semiyearly."

"Biennial" means "every two years."

Most publications use no hyphen for compounds beginning with "bi" or "semi," although it is still sometimes seen.

Boldface type. The title of this entry is an example of boldface. Use this typeface for arrays, for vectors and tensors, and for summation, product, and infinity signs. Prefer it for column heads and totals in tables. Use it effectively for paragraph-level subheads. (See **Headings and subheadings.**) If you are marking a manuscript that will be printed, use a wavy underline to indicate boldface.

Brackets. See **Parentheses and brackets.**

Brand names and trademarks. In scientific papers, avoid using brand names. Instead, use a generic term. If the brand name is necessary for clarity, add it after the first use of the nonproprietary name.

Before transport of the patient, paramedics used an automatic injecting device (LidoPen, Survival Technology Inc., Bethesda, Md.) to administer 400 mg of lidocaine hydrochloride (13.3 per cent, 3 ml) into the deltoid muscle. . . . Immediately before randomization, a monitor-lead electrocardiogram was recorded with an audio cassette recorder (Phillips N2228) in which the audio amplifier was replaced by the CardioBeeper circuitry board (Survival Technology). The electrocardiographic signal was continuously recorded for 60 minutes on a C-120 cassette tape and was reviewed fully on a Lown Trendscriber (American Optical Co., Waltham, Mass.) at the study center.[14]

It is common practice when discussing drugs to give both generic name and brand name. *Administered was lovastatin (Mevacor).* Generic terms like lovastatin require no capitals. Nouns after brand names are not capitalized (*IBM computer, Perkin-Elmer spectrometer*). See also **Eponyms.** The trademark sign is not used in the scientific press.

Bullets. Bullets are circles or heavy dots used to display items in a list. Their size is usually specified in points. • = 6 pt. bullet; ● = 8 pt. bullet. See also **Lists.**

Businessman, businesswoman. But *small-business man, small-business woman.* See **Hyphen.**

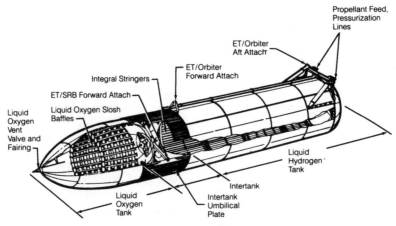

Fig. D-1 Callout.

C

Callout. The callout (words describing parts of an illustration) may be written directly on the illustration and connected with leader lines or arrows to the parts. (Fig.D-1) If the item is complicated, the callout may be handled with numbers or letters on the illustration accompanied by a separate legend or index.

Capitalization. Use initial capitals for proper nouns. Words derived from proper nouns are usually not capitalized (*mendelian, roman numerals*). Do not capitalize initial letters of common nouns after brand names or eponyms (*Hewlett-Packard printer, Reyes' syndrome, Raman spectroscopy*).

Capitalize initial letter in "figure," "chart," and "table" when you refer to a numbered figure, chart, or table: *in Figure 2*.

Use initial capitals for all words in titles and headings except for definite and indefinite articles (*a, an, the*), short prepositions (*at, by, of, in, on, to, vs.*), and conjunctions (*and, but, for, or, nor, so, yet*): *Aspirin, Sulfinpyrazone, or Both in Unstable Angina: Results of a Canadian Multicenter Trial.* Use initial capitals for the first word and for "to" in infinitives.

Rules for capitalizing initial letters of hyphenated compounds in titles vary. In general, capitalize equal elements in compound words ("Epithelial-Cell" in *Effect of Added Dietary Calcium on Colonic Epithelial-Cell Proliferation in Subjects at High Risk for Familial Colonic Cancer;* "Base-Line" in *Table 4. Base-Line Characteristics of the Four Groups of Patients*). Capitalize main elements in premodifying compounds used in titles: "Blood-Filled" in *Figure 3. Sectioned Surfaces of the Spleen, Showing Numerous Blood-Filled Vascular Spaces.*

Most publications do not capitalize the second element of hyphenated compounds used in titles when both parts are necessary to compose the word: *X-ray, Forty-second Trial, Runner-up, Follow-up Study, Self-evaluation Test* but also *Self-Evaluation Test.*

Do not capitalize units of measure in titles or headings (*25,000-dalton Antigen*) unless they appear as words: *Hours of Service vs. Performance.*

Captions. Captions are the words identifying and explaining figures. They are also called *legends*. Figures that are clear to their creators are sometimes incomprehensible to readers who rely on the caption to identify and explain an otherwise obscure illustration.

Use the caption to explain all abbreviations, symbols, and pointers (arrows) used in the illustration:

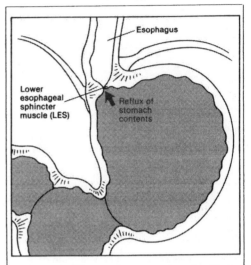

Fig. 1. Normally the lower esophageal sphincter muscle (LES) contracts and keeps the entrance to the stomach from the esophagus closed. Arrow shows reflux of stomach contents. The cause of heartburn is a temporary weakness of the LES.

Use successive arabic numerals for figures. Avoid the superfluous "a photograph of," "a line drawing illustrating," or "a graph showing."

In complicated figures divided into parts with numbers or letters, account for the division in the caption: *Figure 1. To Repair Problem Doors: (a) oil hinge (b) raise pin and add more oil.*

Captions for graphs should include each axis clearly labeled with the quantity measured and the units of measure. Avoid abbreviations if you use no caption other than the axes on the graph.

Caret. Carets are symbols directing the compositor or printer to insert additional information. In manuscript, the addition is shown immediately above the line. In galleys, the addition is shown in the margin. See **Copy editing marks and codes.**

Cautions and warnings. Cautions alert readers to possible harm to equipment. Warnings alert readers to possible danger to themselves. Set off cautions and warnings typographically, using an eye-catching combination of darker or larger type with underlining or a box. Do not divide cautions or warnings that fall at the end of a page. Cautions and warnings should stand as single, striking visual units.

Celsius. See **Degree symbol.**

Cement, concrete. Cement is an adhesive; concrete is a hard building material made by combining an adhesive, such as portland cement, with water and an aggregate such as pebbles and sand.

Charts. See **Tables.**

Citations. See **References, in text** and **References, lists of.**

Cliches. Worn or overused expressions: *state-of-the-art model.*

Co-author. A noun, not a verb.

Coherence. See **Paragraph.**

Collective nouns. A collective noun refers to a group (*series, variety, couple, audience*). When members of the group are viewed as a unit, use singular verbs and pronouns to refer to the group. When members of the group are viewed individually rather than as a unit, use plural verbs and plural pronouns: *The series demonstrates important features of spectroscopic perturbation* (the series viewed as a unit, not individually) but *The series of spectra were transferred onto transparencies using a laser printer* (the series viewed individually). *The majority is in favor* but *The majority of women working at video terminal displays were tested for radiation exposure.* "None" is increasingly construed as a plural. Prefer "not one" for the singular.

Units of measure are regarded as collective singulars and therefore take singular verbs: *30 mL was added.* In general, plurals of quantity and extent take a singular verb so long as they are viewed as a unit: *Twenty percent is a good response rate. Fourteen dollars is the unit price.*

Colon. Use a colon to introduce a list. *The rank order of activity from the most to the least active agent was as follows: ceftizoxime (MIC90 = 0.015 µg/mL); ceftriaxone (MIC90 = 0.06 µg/mL); norfloxacin (MIC90 = 0.125 µg/mL); imipenem (MIC90 = 0.5 µg/mL).*

Use a complete statement before the colon. Do not insert a colon between a verb and object, preposition and object, or verb and complement. Not *Patients received: one of four possible treatment regimes, aspirin (325 mg four times daily), sulfinpyrazone (200 mg four times daily), both, or neither.* But *Patients received one of four possible treatment regimens: aspirin (325 mg four times daily), sulfinpyrazone (200 mg four times daily), both, or neither.*

Use a colon to introduce a restatement or explanation. Here the colon links two sentences when the second sentence explains the first. One can imagine the words "namely" or "as follows" introducing the restatement. The initial letter of the second sentence may be capitalized for emphasis. *Thomas Edison had an inviolate procedure in keeping his laboratory notebooks: He made sure all entries were signed, dated, and properly witnessed. The Nobel prize system has a serious flaw: Former winners nominate candidates, a procedure that makes laureates the targets of many ambitious fellow scientists.*

Use a colon to introduce a quotation. *The editor of the journal had a favorite saying: We do not value what we cannot understand.*

Use a colon to represent a ratio: *The two reactants are consumed in a ratio of 1:2.*

Column headings. Column headings usually include the unit measures, unless these are in the stub (the left column). To make column heads visually distinct, many publications set them in boldface.

Show levels within a column heading either by braces or a straddle rule.

Net Primary Production, g/m^2	
Per Year	Per Day

← straddle rule

Use abbreviations when space is limited, and expand them in the footnotes. Use initial capitals for major words, lowercase for articles, short prepositions, and coordinate conjunctions. See also **Tables.**

Comma. Use commas to set off introductory subordinate clauses and adverbial phrases: *When expressed in* Escherichia coli, *these clones can elicit bioluminescence that is readily visible. Because of the different colors, these clones may be useful in experiments in which multiple reporter genes are needed.*

Upon addition of luciferin to the media, expression of bioluminescence from the tac *vector yielded sufficient intensity to allow measurement of the spectral distribution from intact cells.*

If the adverbial phrase is short—fewer than five words is the usual limit—the comma is not necessary unless the sentence would otherwise be unclear. The comma is used to prevent an initial misreading of the sentence. *To succeed, an experimenter must work hard.* (Without an introductory comma, the sentence might be misread.)

Use commas to set off parenthetical words and phrases (*for example, however, therefore, consequently, in contrast, i.e., e.g.*) or parenthetical (nonrestrictive) subordinate clauses, participial phrases, or identifying nouns. *Her argument was presented clearly; however, I did not agree with the conclusions.* ("However" is parenthetical.) *Increased cell proliferation in the crypt epithelium, accompanied by accelerated attrition of maturing epithelial cells near the mouth of the crypt, would inevitably produce a larger concentration of immature cells near the crypt mouth.* (The "accompanied by" phrase is parenthetical.) *My husband, George, is an engineer.* (The speaker has only one husband. Therefore the name "George" is parenthetical or nonrestrictive and requires commas. Compare with "My brother Bart is an engineer, but my brother George is an accountant.") *The procedures, which were dated, were eliminated from the protocol.* ("Which were dated" is parenthetical, and is set off by commas. Compare with "The procedures that were dated were eliminated from the protocol.")

Use commas to connect independent clauses joined by "and," "but," "for," "or," "nor," "so," "yet." *The total number of epithelial cells per crypt column increased, and the number of labeled epithelial cells per crypt col-* *umn decreased.* If the clauses are short, this comma may be dropped.

Use commas to separate three or more items in a series, whether the items are words, phrases, or clauses. The conjunction before the last item may be preceded by a comma (the *serial* comma). Most academic and technical publications prefer the serial comma. *Neuroendocrine, neuropharmacological, and behavioral studies suggest that dopaminergic systems modulate endogenous opioid system activity.*

Use commas to separate coordinate adjectives (adjectives that could be linked logically by "and"): *early, successful treatment with intravenous streptokinase* but *early warning system.*

Use commas to separate elements in addresses and dates: *We sent the assay to Biologic Labs, Ontario, Canada, for processing.* See **Dates, punctuation of.**

Use commas to set off indirect and direct quotations: *The question is, Who will pay for the project if other sources fail? He asked, "Who will pay for this project?"*

Use commas to set off degrees and titles that appear after a name: *John Reiser, Vice-President, chaired the meeting* but *Vice-President Reiser chaired the meeting.*

Many publications use commas to separate groups of thousands in numbers beginning with either four digits or numbers larger than 9999. An increasing number of publications, however, follow the International System of Units (SI) standard, in which digits are separated into groups of three by a small space instead of a comma, as some countries use the comma as a decimal marker. See **SI units.**

Compositor. Typesetter; person who, in process of constructing tables and setting type, enters the changes, corrections, and codes placed on the manuscript by the copy editor. Notes intended for the compositor are circled in the margin. For instance, to indicate that a numbered list should be set according to specifications agreed to by the publisher and the compositor, the copy editor writes and circles NL in the margin next to the list. See also **Copy editing marks and codes.**

Compound subject. Compound subjects (subjects joined by "and") are usually plural unless thought of as one: *Research and development is one of the company's major assets.*

Compound words and modifiers. Compound words, formed from the combination of two words or more, are an important source of new vocabulary in English. As they first gain currency, they may appear spelled either solid (*workstation*) or open (*work station*). They may also appear hyphenated (*work-station*). As they begin to establish themselves in the language, they are as likely to appear open (*word processor, earth works, hard hat*) as solid (*wordwrap, earthquake, hardwood*). In general, the progression in U.S. English is from open to solid as a compound becomes established as a permanent lexical item.

Established compounds are in current dictionaries. For recent coinages, check with a professional society establishing conventions for emerging expressions in your field.

Some compounds that are usually hyphenated include noun compounds in which the second base is an adverbial preposition (*runner-up*), coordinated compounds (*writer-editor*), compounds expressing numerals and fractions (*twenty-three*), and compounds in which the first base is a single capital letter (*O-rings*).

If you use an adjective with a solid compound (*businessman*), you may need to introduce a hyphen and a space to avoid confusion. *small-business man.* See **Hyphen.**

Compound modifiers are compounds used as adjectives, and include noun-participles (*tissue-bound substance, far-reaching vision*), adjective-adjective (*blue-gray mix*), and noun-adjective (*rock-solid substance*).

Capitalize major elements in compound modifiers used in titles, subtitles, and headings: *Two-Photon Spectroscopy. Low-Yield Radioactive Waste.* See also **Capitalization** and **Hyphen.**

Comprise. The whole comprises its parts. *The review volume comprises ten articles* not *Ten articles comprise the review volume.*

Continual vs. continuous. Use "continual" for that which is repetitive, but interrupted. *Continual testing indicated an unusually high failure rate.* Use "continuous" for that which is uninterrupted. *A laser that continuously emits light is a continuous-wave(cw) laser.*

Continuity. See **Paragraph** and **Transitions.**

Convince, persuade. Convince "that" or "of." Persuade "to."

Copy. Typescript (usually double-spaced) and original artwork.

Copy editing marks and codes. Most copy editing marks go on or above the lines of manuscript, which are typed double-spaced to allow room for the marks. (All proofreading marks, in contrast, go in the *margins* of single-spaced, typeset proof.)

correct	the visible spectrum
delete	30 mL
delete and close	pyrotechnic compositions
flush left	[table column
flush right	date]
center	] heading [
insert space	scientific goals
close up	where as
italics	Drosophila melanogaster
boldface	Introduction
roman	set in roman → *hypovirulence in C. parasitica*
bullet	● (7 pt.)
en dash	non fossil-energy sources, 30-40 units, 1998-99
em dash	Table 18-Incidence of Melanoma
superscript	32,768 (or 2 15); a Z0 particle
subscript	CO2
apostrophe	time's arrow
comma	Global climate warming some claim threatens the planet.
period	about
semicolon	
colon	There are two classes of fermions: leptons and quarks.
paragraph	¶ Toyota, Nissan, and Mazda are already selling cars with navigation systems in Japan at premiums of up to $4,000. Combined sales are estimated

Some copy editing marks are codes intended for the typesetter or compositor. The compositor matches the code (sometimes called a "macro"), which is circled and placed in the margin by the copy editor, to such previously agreed-to specifications as trim, margins, paragraph indents, type size, typeface, leading, color, and position. For instance, if the copy editor writes BL next to a list, the compositor knows that according to the agreement with the publisher the list must be set in 10/12 Times Roman × 30 picas, that it should be set justified, that all lines should be indented 18 pts. from left and right margin, and that medium bullets should hang to the left of the list.

Codes are usually simple mnemonics; the following codes are representative, although they vary from company to company. All codes are circled; the circle means that the information should not be set in type.

change words	to be about 1,000 units a month. The required ~~luminosity~~ *beam intensity*
insert words	their amplitudes *oscillation* particle beams that
let stand (ignore marks)	are ~~almost~~ impervious
use other form (fig. vs. words)	The (three hundred) *300* page centennial version is (15-fold) *fifteen-fold* longer than the first version. The cross-index lists about (eighty) (thousand) *80,000* names. [For clarity, write out changes if capitalization or punctuation added.]
hyphen	satellite=based tracing system
insert hyphen	the 64K protein
insert parentheses	(See Figure 1)
plus or minus	w+/ w pairs
capitals	30 ml
italic capitals	See Figure 1.
lowercase (one letter)	John Von Neumann
lowercase (series)	FIGURE 1—RADIOACTIVELY LABELED DSRNA
indent 1 em	Table Entry
indent 2 ems	Table Entry
transpose	represents (a more than) 20 percent increase
Greek	a and b particles
move	. Indentation or . Change in size or darkness of type
align	The total number of epithelial cells per crypt column increased, and the number of labeled epithelial cells per crypt column decreased.
use regular type, not boldface	See **SI units** ROM

chapter number	1 (CN)
chapter title	The Composition of Matter (CT)
first-level heading	Demographic (1) Analysis
second-level heading	Fertility Rates (2)
third-level heading	Aging of the (3) population
fourth-level heading	Marriage and divorce (4)
set in two columns	Major demographic and sociological changes directly influencing family composition have taken place in this century, with the pace of change accelerating in the past two decades. (2-COL)
numbered list	1. Hold screen vertically with legs down. 2. Pull legs outward. 3. Place tripod on ground. (NL)
bulleted list	• **Shopping.** Pick up perishables last. • **Refrigeration.** Leave products in store wrap. (BL)

italicized list

Proposing what you
want to do
Reporting what you
did
Instructing ⌍ (IL)
● = 6 pt. bullet
↖ *comp: set 6 pt bullet*

comments for
compositor

equation $y = v + x$ ⌐ (EQ)

table ⓣ

table number (TN)

table title (TT)

table column
heads (TCH)

table footnotes (TF)

(TN) (**Table 3.** Marriage and Divorce Rates in 10
Countries, Selected Years, 1960–1986 (TT)

Country	1960	1970	1980	1986
Marriage rates (per 1,000 population, ages 15 to 64) (TCH)				
United States	14.1	17.0	[1]15.9	15.1
Canada	12.4	14.3	11.8	10.2
Japan	14.5	14.4	9.8	8.6
Denmark	12.2	11.5	7.9	9.0
France	11.3	12.4	9.7	7.3
Germany	13.9	11.5	8.9	8.7
Italy	11.7	11.3	8.7	7.5
Netherlands	12.7	15.2	9.6	8.7
Sweden	10.2	8.2	7.1	7.2
United Kingdom	11.5	13.5	11.6	10.6
Divorce rates (per 1,000 married women) (TCH)				
United States	9.2	14.9	22.6	21.2
Canada	1.8	6.3	10.9	12.9
Japan	3.6	3.9	4.8	5.4
Denmark	5.9	7.6	11.2	12.8
France	2.9	3.3	6.3	8.5
Germany	3.6	5.1	6.1	8.3
Italy	[2]	1.3	.8	1.1
Netherlands	2.2	3.3	7.5	8.7
Sweden	5.0	6.8	11.4	11.7
United Kingdom	2.0	4.7	12.0	12.9

ⓣ

[1]Beginning in 1980, includes unlicensed marriages registered in
California.
[2]Not available.
Sources: Statistical Office of the European Communities, *Demographic Statistics, 1988;* and various national sources. (TF)

long quotes

Before transport of
the patient,
paramedics used.

descriptive
comments
on text

an automatic
injecting device to
administer 400 mg
of lidocaine
hydrochloride into
the deltoid muscle. ⌐ or (Q)

Definition inserted
for clarity. ⌐ (DES)

example

Objectives: (i)
Develop a model.
(ii) Determine the
factors prohibiting
throughput. ⌐ (EX)

notes

(Sentence 1 has a
subject-verb
agreement error;
Sentence 2 has a
mistake in parallel
elements; Sentence
3 has a missing
hyphen.) ⌐ (NO)

abstract

Abstract: This study
evaluates the
effects of. . . . (AB)

epigraphs

My primary concern
is to render my
work as clear and
intelligible as it is
true. ⌐ (EPI)

Copyfitting. Estimating the space needed for
a document at a given type size.

Crisis, crises. See **Greek and Latin words
and expressions.**

Criterion, criteria. See **Greek and Latin
words and expression.**

Cropping. Removing extraneous parts of illustrations. Photographs may be cropped or edited by marking areas to be deleted with a special marker on the edges of the photograph. Do
not mark the photograph directly. If you do not
have a cropping pen, use a transparent overlay
and make crop marks on it.

Do not crop graphs unless essential. Cropping saves space, but it also produces distor-
tions. If you must crop, use a zigzag (Z) to
indicate the deletion.

D

Dangling modifiers. See **Modifiers, misplaced or dangling.**

Dashes. The three dashes are the hyphen (-), the en dash (–), and the em dash (—). The en dash, which is longer than a hyphen when set in type, provides a useful distinction when you want to add a hyphenated modifier to a term that is already a compound: *nonidentical–two-photon excitation*. Here "nonidentical" modifies not "two" but "two-photon." The en dash allows us to see this relational distinction.

Use the en dash for ranges: *1990–99*.

Use the em dash (represented on a typewriter by two hyphens) instead of parentheses for asides, definitions, or other interruptions, when you want a more emphatic effect. Compare *Integrated circuits—slices of pure silicon embedded with traces of impurities—are used in the new process* with *Integrated circuits (slices of pure silicon embedded with traces of impurities) are used in the new process.*

Use the em dash for table designators: *Table 1—Incidence of Melanoma.*

Do not combine the em dash with a comma, semicolon, or colon. Not *"The opposition to cold fusion,—I use the term "opposition" loosely,—has brought together many diverse parties"* but *"The opposition to cold fusion—I use the term "opposition" loosely—has brought together many diverse parties."*

Prefer commas to em dashes to separate phrases and nonrestrictive information. Do not use the em dash when the comma is adequate. In this example, the em dash is misused for the comma: *Tables enable you to present certain materials—especially numbers—more clearly than you can in text.*

Data. While "data" is increasingly construed as a singular in informal discourse (*This data is . . .*), the plural is firmly established for formal use (*These data are contradicted by later findings*) unless you speak of one fact or datum, in which case you may want to say "one item of data." See also **Greek and Latin words and expressions.**

Dates, punctuation of. No comma is needed to state month and year only: *We ordered a pulsed dye laser in October 1990 and a pulse compressor in January 1991.* If you specify a day, separate day from year by comma, and always separate year from any remaining text in the sentence. *The earthquake of October 17, 1989, measured 7.1 on the Richter scale.*

If you use the European system of day-month-year, no commas are needed. *We ordered the pulsed dye laser on 22 October 1990 and the pulsed compressor 22 January 1991.* Avoid all-numeral representation of dates (*3/18/91*) in text, but if you must use this form, be careful to give the month first, not the day. (Therefore *March 8, 1991* becomes *3/8/91*, not *8/3/91*.)

Decimals. Use figures for decimals: not *five-tenths of a gram* but *0.5 g*. Use decimals with a unit of measure: *5.25 g*. Use a zero before the point: *0.25 g*. Decimals represent precision. Do not, therefore, add zeros to the right simply to balance the appearance of a table.

Use decimals instead of fractions to indicate precise measurement: *7.25 cm* not *7¼ cm.*

Some countries use the comma, rather than a point, as a decimal marker. See **Comma** and **SI units.**

Definitions of terms. Set off definitions with parentheses, dashes, or an expression like "defined as." If the technical term is not used again, put it in parentheses, and use a more general term in the text. *The topic of pathogens that are transmitted from animals to human beings (zoonoses) has received attention recently.* See also **Glossary, Italics,** and **Jargon.**

Degree symbol. For angles, do not space between degree symbol and number: $90°$. For the Celsius and Fahrenheit temperature scales, some publications use a space after the number, but no space between the degree symbol and the C or F. No period follows: $6°$ C temperature. Other publications use no space between degree symbols and number: $6°$ F temperature. No degree symbol is used with the kelvin temperature scale: 0 K temperature.

Descenders. Descenders are the parts of letters like *p, q,* and *y* that dip below the print line.

Descriptors. See **Key words.**

Different from vs. different than. Use "different from."

Dimensions and ranges. To prevent misreading, repeat the word "million" in expressions like *$4 million to $5 million* (not *$4 to $5 million*). If no misreading is anticipated, use only the final unit of measure (*2–5 mg*). Use the en dash with ranges. Use the successive hyphen in expressions like *2-, 3-, and 4-day trials* or *5- to*

7-cm diameters. See also **Hyphen.** Render dimensions in words only when a sole dimension is given: *A five-pound weight.*

Discussion. See **Results and discussion.**

Due to. Use "due to" to modify nouns and pronouns, not verbs. See **Because of.**

E

Each other, one another. For two items, use "each other"; for more than two items, use "one another." "Each other" and "one another," like "either," "everyone," and "anybody," are singular, and therefore take singular verbs and singular pronouns. See **Gender** and **Number of subject and verb.**

Effect, affect. See **Affect, effect.**

e.g. vs. i.e. The common error is to mistake e.g. (meaning *exempli gratia*, "for example") for i.e. (meaning *id est*, "that is to say."). Both expressions are parenthetic and set off by commas.

Either. "Either" takes a singular verb unless it is paired with "or." Then the verb agrees with the nearer subject. *Either the original article or abstracts of it are available. Either abstracts or the original article is available.* Note that "either" is paired with "or," not "nor."

Ellipses. In text, use three points (. . .) to represent deleted material within a sentence, four points for omissions at the end of a sentence. Use a space before and after if the ellipses are within a sentence: (*Medewar comments that "a lecturer can be a bore . . . because he goes into quite unnecessary details about matters of technique."* Indicate changes in capitalization with brackets: *As Medewar comments, "[P]eople with anything to say can usually say it briefly; only a speaker with nothing to say goes on and on as if he were laying down a smoke screen."* In tables, use points (or dashes) if no data are available, or if the category is irrelevant. (Zeros or ciphers stand for a measurement of zero, not for a statement that "no measurements were made" or "this category is not relevant.")

Enormity, enormousness. "Enormity" means "horror"; "enormousness" refers to size. Not

The enormity of the lasers used in the fusion experiments startled the visitors to the laboratory but *The enormousness of the lasers used in the fusion experiments startled the visitors to the laboratory.*

Ensure. See **Insure vs. ensure, assure.**

Enumerations. See **Lists.**

Eponyms. Eponyms (names based on persons or places) are frequently used in science and technology to describe an instrument (*Bunsen burner*), a field of study (*Raman spectroscopy*), a phenomenon (*the Einstein-Podolsky-Rosen paradox*), a disease (*Reyes' syndrome*), or a test, method, or procedure (*the Stockbarger method*). Capitalize the eponym, not the modifiers: *Bohr magneton, Lorentz-polarization, Planck constant, extended Huckel, Born-Oppenheimer approximation, Benedict-Webb-Rubin equation.*

The possessive is used with most eponyms (*Avogadro's number*), but the usage is declining (*Lewis acid*). For double and triple names, no possessive occurs (*Einstein-Podolsky-Rosen paradox*).

et al. Use no comma in this expression unless you name a series. *Meyerson et al.* but *Meyerson, Garetz, Lombardi, et al.*

etc. Do not use "etc." with "and." Not *Meyerson, Garetz, Lombardi, and etc.* Do not use the redundant *etc.* when introduced by expressions like "such as" or "for example." Not *Light waves are scattered by obstacles such as dust particles, gas molecules, etc.* but *Light waves are scattered by obstacles such as dust particles and gas molecules.*

Everyone, every one; anyone, any one. Use as separate words to emphasize each member of the group. *Every one of the department's five specialists in organometallic chemistry presented results* but *Everyone attended the seminar.*

"Everyone" and "anyone" are singular, and therefore take singular verbs and singular pronouns. See **Gender** and **Number of subject and verb.**

Ex-. This prefix is usually hyphenated, but not in expressions such as *ex post facto* and *ex parte.*

Exdent. To begin a line of type outside the margin.

The first letter of this sentence is exdented.

Exploded view. Close-up illustration of parts of the whole, often shown in the order of assembly.

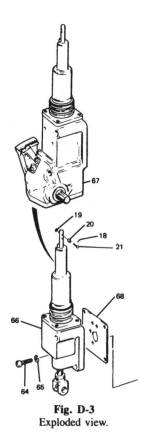

Fig. D-3
Exploded view.

F

Fahrenheit scale. See **Degree symbol.**

Fewer, less than. The traditional distinction is to use "fewer" for that which is individual and countable (*Fewer instruments are available. Do not use this column if you have 10 responses or fewer*), and "less than" for amounts (*We have less liquid nitrogen on hand than expected*). The distinction is not observed in mathematical expressions such as "equal to or less than."

Figures. A figure is an illustration such as a photograph, line drawing, graph, or chart.

Number figures with successive, arabic numerals, designate them (either *Figure 1* or *Fig. 1* in the caption), and use the designation on first reference in the text ("In Figure 1 . . . See *Figure 1*). Do not say "the above figure" if you do not control placement of the figure, as it may appear in print before, beside, or after the reference. If you do control placement of figures, put them as soon as possible after first mention in text. To avoid confusing readers, do not place figures before first mention. Any figure displayed should be cited in the text.

Spell out abbreviations and acronyms used in figures. If you use a graph without an accompanying caption, spell out abbreviations.

Explain any arrows, pointers, or special symbols used in figures in the caption. When space permits, use graphs to illustrate trends and tables to emphasize specific numbers. Prefer line drawings to show close-ups, to show the parts of the whole, to illustrate the function of a mechanism, and to reduce distracting detail. Cutaways and cross sections permit the reader to see beneath the surface. Schematics and sketches allow the reader to focus on one aspect of a complex design.

Photographs are useful to orient the reader before moving to line drawings (a photograph of the coastline followed by a navigational map, a photograph of a hydraulic governor followed by outline drawings, schematics, and exploded views to show detail). Photographs are useful to record an event (the collapse of a freeway during an earthquake, followed by line drawings to explain the causes of the collapse). If you are providing photographs, supply good prints with sharp contrasts. Label each figure on the back, numbering consecutively in the order of reference in the text. See also **Abbreviations, Captions, Cropping, Overlay,** and **Tables.**

First person. From the Renaissance to the early twentieth century, "we" and "I" were entirely acceptable words in scientific discourse. In 1615, Harvey wrote in *An Anatomical Disquisition on the Motion of the Heart-Blood in Animals,* "I finally saw that the blood, forced by the action of the left ventricle into the arteries, was distributed to the body at large and its several parts, in the same manner as it is sent through the lungs, impelled by the right ventricle into the right pulmonary artery, and that it then passed through the veins and along the vena cava, and so round to the left ventricle in the manner already indicated, which motion we may be allowed to call circular." In 1869, first person was still alive and well when Lister reported to the British Medical Society that, "In the course of an extended investigation into the nature of inflammation, and the healthy and morbid conditions of the blood in relation to it, I arrived, several years ago, at the conclusion that the essential cause of suppuration in wounds is decomposition."

In the early twentieth century, first person fell out of favor among many people who apparently believed that they would sound more objective if they did not include references to themselves in scientific accounts. The result was a convoluted style that quickly generated its own passive cliches (*it was observed, it was found that, it was demonstrated that*). Recently many publications have begun to fight back, urging their readers to use first person when appropriate, particularly in the introduction and conclusion to a paper, where the author's voice is most clearly heard discussing and interpreting. First person helps writers avoid passive cliches, specify who did what when this information matters, and achieve a more direct tone. It is often the most succinct, readable way to cast sentences. See also **Active vs. passive voice** and **Scientific papers.**

Flammable vs. inflammable. *Flammable* and *inflammable* both mean "easily ignitable." Prefer *flammable* as some people think "inflammable" means "nonflammable" or "noncombustible."

Flush left, flush right. To align copy vertically along the left (flush left) or right (flush right) margin. See also **Justified right vs. ragged right.**

Fog index. See **Readability indexes.**

-fold. One-word numbers used with this suffix are usually written out and are usually unhyphenated (*twofold* but also seen *two-fold*). When the numbers are already hyphenated (*sixty-four*), use numerals and hyphens (*64-fold*).

Folio. Page number. Usually the page number occurs after the running head at the top of the page. A **drop folio** is a number printed at the foot of the page.

Font. A complete set of letters, numbers, and symbols of a typeface design or face (for instance, Helvetica) that are the same size. **This font is Helvetica 14 pt. bold.**

Footnotes. In general, try to include material within the text, rather than using separate footnotes. Traditional use of footnotes for the first page of scientific articles includes separate listing for date of submission and acceptance, affiliation of authors, notice of death of an author, disclaimers, and address for reprints or further information.

Use footnotes beneath tables for information that will not fit into the title or column heads. The traditional order of footnotes is, in superscript, asterisk (*), dagger (†), double dagger (‡), section mark (§), and parallels (‖).

Indent before using the symbol.

Use footnotes beneath tables for statistical significance, to explain units of measure, to identify abbreviations used in column heads, to acknowledge any references or sources, and to provide explanations beyond those given in the text. Place footnotes at the bottom of the table, not the bottom of the page.

Foreign words and phrases. Initially shown in italics, many foreign words and phrases soon appear in ordinary (roman) print when they become popular. "In vivo," "in vitro," "in utero," and "in situ," for example, are no longer italicized in many publications. Accent marks also tend to disappear as phrases become popular ("cliche").

Do not hyphenate foreign words and phrases used as compound modifiers ("in vivo process," "ad hoc measure"). For plurals, see **Greek and Latin words and expressions.**

Format. A printer's term for the general appearance of the document. If a copy editor is working for a publisher, the format will have been decided in advance. When the copy editor then checks, for instance, tables submitted with the manuscript, the copy editor uses codes to mark the tables (e.g., TFN for footnotes, and TCH for table column heads). The compositor then uses these codes or macros as a guide to type size, typeface, line position, leading, and color, among other specifications. See **Copy editing marks and codes.**

Copy editors working with desktop publishing use the format commands linked to their software.

Those copy editors working without specifications will want to develop their own format sheet. Such a format sheet might, for instance, specify line position for running head, folio, first-, second-, and third-level headings, bullets, margins, and indentation.

Here are some typical categories to consider in developing a format sheet.

TITLE

All caps? Initial caps only? Centered? Double-spaced before and after? Other: Underscored or boldface?

Section Headings

On line by self? Begin flush left? Begin line position #?
All caps? Initial caps? Underscore? Boldface? Double space before and after? Other:

First-Level Headings

On line by self? Begin flush left? Begin line position #?
All caps? Initial caps? Underscore? Boldface?

Second-Level Headings

On line by self or text follows on same line? All caps or initial caps?

Lists

Bulleted
Size of bullet? Space between bullet and text?
Begin line position #? Line position for continuation of items?
Numbered
Enclose numbers with parentheses? Use periods?
Begin line position #? Line position for continuation of items?
Punctuation
For all lists, including simple ones?
For lists with punctuation within?
For any item with a verb?
Hanging text?
Begin continuation at line position #?

Body

Typeface, size, width, justification, and leading for narrative?

For quoted material?
For examples?
For figure callouts?
For captions?
Margins: Head? Foot? Gutter? Outside?
Tab sets? Space between paragraphs? Paragraph indents?
Page numbering? Consecutive? By section? Running heads?

Fractions. If the fraction appears with a whole number ($3\frac{1}{3}$), use numerals unless it begins a sentence. If the fraction appears without a whole number, use words: *one-third of the respondents, one-twelfth of the sample.* Also seen: *one third of the respondents* but never $\frac{1}{3}$ *of the respondents.* Hyphenate fractions that are compound modifiers (*seven-eights majority*). As decimals represent more precise measurements, do not use them in place of fractions if doing so suggests an exaggerated degree of accuracy.

G

Galleys. See **Proof.**

Gender. While English offers a third-person plural pronoun that gracefully includes both masculine and feminine (*they, them*), it lacks such a pronoun for third-person singular (*he, she, him, her*). This leads to a problem when the writer uses a term like "everyone," "everybody," "each person," or "one," and then casts about for the proper singular pronoun to accompany it. The historical solution has been to use a masculine pronoun, thereby excluding females: *Everyone should submit his laboratory notebook for indexing every six months.* One solution to this exclusionary language is to change singular antecedents to plurals: *Staff members should submit their laboratory notebooks for indexing.* Some writers recast the sentence: *Please submit all notebooks for indexing every six months.*

The occasional use of "he or she" is also tolerated, although many writers and readers find it cumbersome. Avoid "he/she."

Generic names. Prefer generic terms unless the brand-name product differs from the generic one, or giving a brand name will help readers understand an argument or replicate a procedure. See **Brand names and trademarks.**

Genetics, terminology for. See **Nomenclature.**

Genus names. The genus name is first, using an initial capital, followed by the species name, using no initial capital: *Toxoplasma gondii*. Lowercase is retained for the species, even in a title.

Use italics for genus and species, so long as the genus is singular. Use an abbreviated form of the genus, but not of the species, for subsequent references:

Nearly all our knowledge of beetle luciferase is derived from studies of a single species, the North American firefly *Photinus pyralis*. . . . Recently we cloned a cDNA that codes for the luciferase of *P. pyralis*.

A period may or may not follow the abbreviated genus name (usually *E coli* but also widely *E. coli*).

The genus name is not capitalized in the plural or as an adjective (*streptococcal*) unless it is used at the beginning of a sentence or in a title.

The *International Journal of Systematic Bacteriology* publishes guidelines for the names of genera and species. See also **Nomenclature.**

Glossary. A glossary is an alphabetized list of terms and their definitions. Use etymological information (*bit, from binary digit*), examples, comparisons, contrasts, and figures as necessary to expand definitions so that they will be accessible to readers.

Graduate. People may either "graduate from" or "be graduated from." Avoid *Mr. Sinclair graduated MIT in 1991.*

Graphs. See **Figures.**

Greater than or more than vs. over. Use "greater than" or "more than." The term "over" gives little sense of quantity. Not *Over 100 people tried the sample* but *More than 100 people tried the sample.*

Greek and Latin words and expressions. For plurals, the trend is to prefer the English (anglicized) version. Thus *appendixes* is supplanting *appendices, craniums* is overtaking *crania,* and *formulas* is replacing *formulae.* Still, many Greek and Latin plurals remain, among them *algae, criteria, spectra,* *phenomena, crises, bases, axes, hypotheses, media, strata,* and *data.*

Greek letters. When Greek letters are used in hyphenated compounds (β-adrenoreceptors) in titles, use initial capitals for the word following the Greek letter: *β-Adrenoreceptor and α-Adrenoreceptor Block Properties in Labetalol.*

If you are preparing a manuscript and do not have Greek type to indicate ν, κ, μ, and η, use the roman v, k, u, and n, but circle each on first use and explain in the margin that Greek letters should be inserted.

Group nouns. See **Collective nouns.**

Gutter. The inner margin between the text and the binding.

H

Hairline rule. A thin rule (about 0.5 point).

Hanging paragraph. A useful device to distinguish items in a displayed list, a hanging paragraph indents all lines but the first. This device is also called a "flush and hang":

G. Herzberg, *Spectra of Diatomic Molecules* (Van Nostrand Reinhold, New York, 1950).
W. M. McClain and R. A. Harris, in *Excited States,* Vol. 3, edited by E. C. Lim (Academic, New York, 1978), pp. 1–55.

When information for each item in a displayed list continues past the first line, the continuation usually starts under the first letter of the line above, not the number:

In September 1969, two months after the initial lunar landing, a Space Task Group chaired by the President offered a choice of three long-range plans:
1. An ambitious program for a manned Mars expedition and a 50-person Earth-orbiting station.
2. An intermediate program, costing less than $8 billion annually, that would include a Mars mission.
3. A relatively modest $4 billion program that would include the Space Shuttle.

Headers and footers. See **Running feet and heads.**

Headings and subheadings. Headings and subheadings help readers process information quickly: They allow readers to see divisions in the text, to sort major and minor categories, and to scan selectively.

To do an effective job, headings and subheadings should be informative, distinctive, and consistent. To make them informative, include a succinct preview of the topic that follows. To make them visually distinct, make sure they stand out from the running text. Do this by using a combination of white space (empty carriage returns before and after the headings to give the words emphasis), capitalization, larger type, underlining, line position, and different treatments (boldface, italic). Once you've chosen a style of headings for each level, make headings consistent by using the same typeface, size, and line position for all choices at each level.

Formats for headings and subheadings vary widely. Here is one typical scheme:

SECTION HEADING. Centered, all capitals, bold. Two carriage returns before and after.

SUBHEADING. Centered, all capitals. Carriage return before and after.

Primary Heading. Side, initial capitals, bold. Carriage return before and after.

Secondary Heading. Side, initial capitals, italics. Carriage return before and after.

Indented Heading Underlined, initial capitals. On its own line.

Paragraph Heading. Begins at paragraph opening. Bold. Initial capitals. Closed with point. Followed immediately by text.

Headings in this pattern would look like this:
TITLE
CHAPTER
Section Head
Subsection Head
Indented Division
   Paragraph Heading.

Use initial capitals for all words in major headings except coordinating conjunctions, short prepositions, and definite and indefinite articles that are not the first word. Capitalize the first word and the "to" in infinitives. Capitalize major elements in compound words and compound modifiers: *A 69-Year-Old Man with Peripheral Vascular Disease and Hypersplenism.* Do not capitalize abbreviated units of measure. See **Capitalization.**

Heads. See **Margins.**

Hyphen. Soft hyphens are those inserted by the author to divide a word into two parts between the end of one line of text and the start of the next line. Avoid soft hyphens when possible, but if a division is unavoidable, break at the syllable, after the prefix, or before the suffix. Do not divide contractions, acronyms, abbreviations, or one-syllable words. Do not leave a single letter at the end or beginning of a line. If the word has a hard hyphen, use the hyphen for the break. Do not add a second hyphen if possible.

Hard hyphens are those used to separate the bases of a compound word, or to separate prefixes or suffixes from a base. Some rules for hyphenation vary according to the practices of publications. The following practices are standard.

Use the hyphen in compounds for numbers between 21 and 99 when written as words, except for multiples of ten (*twenty-nine lives, sixty-four questions*); for fractions (*two-thirds vote, five-eighths of a mile*); for compounds beginning with a single capital letter (*O-rings, H-bomb*); for coordinate compounds (*attorney-client relationship, Japanese-English dictionary*); for noun-adverb compounds (*runner-up*); for premodifying expressions for age, weight, time, size, and quantity that are written open (no hyphen) when they are not premodifiers (*an apparatus that is five years old* but *a five-year-old apparatus; a trial that lasted three hours* but *a three-hour trial; the stamp costs 19 cents* but *a 19-cent stamp*); for ad hoc premodifying compounds (*a pay-as-you-go scheme; a signal-to-noise ratio*); and for adjective compounds in which the second base is a participle (*menu-driven, pride-filled, polynomial-based descrambler, tissue-bound substance*), an adjective (*blue-gray mix; rock-hard substance*), or a noun (*solid-state chemist, high-frequency measurement*).

Note that the hyphen is used in ad hoc compounds (*a step-by-step procedure*) and adverb-adjective compounds (*a well-prepared site; an ill-prepared report*) that precede the noun, but is dropped when the compound follows the noun: *The procedure was recounted step by step. The site was well prepared.*

If the first base ends in **ly,** do not hyphenate: *a poorly edited manuscript.*

For prefixes, use hyphens to separate prefix from base if the base begins with a capital or figure (*pre-1990, ultra-German, un-American*); to avoid doubling an **i** or tripling a consonant (*anti-isolationist, wall-like*); to avoid ambiguity (*re-cover a chair* but *recover from illness, re-form an extrusion mold* but *reform a drunkard*); after "ex," "half," "quasi," and "self" (*ex-partner, half-finished, self-analysis*).

American English tends to drop the hyphen after prefixes. Thus *co-operate* becomes *cooperate, co-ordinate, coordinate,* although some editors still mark the prefix with an umlaut: *coöperate.*

Use either a hyphen or an en dash to replace "to" or "through" (*pages 5-15*). Use the hyphen for suspensive constructions: *a 4-, 5-, or 12-stringed musical instrument; pro- and anti-government forces.* Some publications do not permit suspensive hyphens with solid compounds. Not *homo- and heterosexual relationships* but *homosexual and heterosexual relationships.*

Many compound words that are initially formed with a hyphen soon evolve into solid words. In general, there is a progression from open to solid as a given compound establishes itself as a lexical item. See **Compound words and modifiers.**

I

ill-. Only hyphenate "ill" when it is used in an adjective compound before the noun. *An ill-formed solid* but *The solid was ill formed.*

Illustrations. See **Figures.**

Impacts, impacts on, has an impact upon. Prefer "affects" to the still questionable use of "impact" as a verb, or to the nominalized "has an impact upon."

Include. Use "include" to introduce a list that may not be complete. *The authors of the study include Susan Calabresi, M.D., and Kenneth Kazaemik, Ph.D.* See also **Comprise.**

Indent. Indented type begins inside the margin. (Compare with **Exdent,** to begin a line of type outside the margin.) Paragraphs are cus-

tomarily indented five spaces in manuscript. **Indentation devices** are symbols such as bullets or squares used to mark each of the items in a displayed list. See **Bullets, Format,** and **Lists.**

Infinitives. The infinitive is the base form of the verb, preceded by "to": *to analyze.* In a split infinitive, an adverb separates "to" and the base form: *to carefully analyze.* The prohibition against split infinitives is strong and widespread. In general, recast the sentence to avoid splitting an infinitive, unless you are sure the split infinitive is more direct than its reworded alternative.

Inflammable. See **Flammable vs. inflammable.**

Instructions, SOPs (Standard Operating Procedures), and procedures. Instructions and SOPs are typically written by those who know for those who do not. When experts write for beginners, they must be not only methodical, but imaginative. What information will the beginner need? In what order? Which terms need to be defined? Which visuals will help? A common problem is for the expert to write above the head of the beginner, compressing or omitting important details. To avoid this, make sure the reader is oriented to the task. The new user needs a framework before the detail. What is the job? Why is it necessary? What equipment must be on hand? What object, process, or outcome should the user expect? This may mean a brief introduction before the steps or, for longer jobs like manuals, an introduction to the manual plus an introduction to each section of the document.

Place cautions and warnings before, not within or after, instructions for dangerous or hazardous steps. Set off cautions and warnings typographically, using underscore, boxing, boldface, or a combination of these devices. Make wording specific, not vague. See **Cautions and warnings.**

Present material step by step, with one instruction to a step when possible. Use the active voice. Not *The exposed area of the disk should not be touched* but *Do not touch exposed area of the disk.* Do not switch person in the same sequence. Not *First, depress the lever and then the user raises the screen* but *Depress the lever and then raise the screen.*

Prefer the imperative. The writer may sound like a drill sergeant barking out orders, but this is a plus when the user needs to know what to do at each stage.

Prefer positive commands to negative ones. Because negatives are indirect (*does not have* for *lacks; are not the same* for *different*), they are easily misread. Avoid *Do not pull lever forward to increase tension; pull it back.* Use *Pull lever back.* If the negative is essential for clarity or safety, place it after the positive, and underscore the negative in the sentence. Revise *Waste hazardous reagents should not be poured in the sink. Instead, they must be collected in a waste can* to *Collect all waste hazardous reagents in a container. Never pour them down the sink.*

Use tabular format for instructions for absolute beginners. Use narrative format, which places a far larger burden on the reader, for more sophisticated audiences.

Set off instructions with numbers if sequence is important, or bullets when sequence is not crucial. Take advantage of headings, subheadings, bold and underline keystrokes, and generous indentations to help the eye do its job.

Check illustrations to make sure wording used in the callouts, captions, and labels matches wording in the text. Errors accumulate when the writer revises the text, but forgets to adjust the corresponding captions and callout.

Unlike instructions, which are directions written for beginners, procedures and protocols assume a readership with more training within the field. For instance, in a scientific paper or technical report, the procedural section tells what the researcher did in sufficient detail that another trained worker—not a beginner—could repeat the steps and obtain comparable results. Often procedures and protocols are cast in the passive voice (*All laboratory notes should be signed, dated, and witnessed at the end of each day* instead of *Sign, date, and witness all laboratory notes at the end of each day*), as they are written for submission to a regulatory agency.

Insure vs. ensure, assure. *Insure* and *ensure* both mean "to make sure or certain." *Insure* is usual for information related to insurance; *ensure* is usual in all other contexts. "Assure" means "to promise." *We assured him that the procedure could be completed quickly.*

Introduction. In a technical or scientific report, the introduction gives the background. It answers these questions: What is the problem? Why is the problem significant? Who else has studied this problem? How does this work add to or differ from related work? If the author is reporting a method, why is this method preferable to its precursors? What are the scope and limits of this report?

If the report is short, a simple statement either of objective or of objective, scope, and limits may be adequate.

For some reports, a brief statement of the relationship of the reported work to other work in the same field suffices. For longer, more scholarly reports, the author must demonstrate a thorough knowledge of related work in the field by citing pertinent literature.

Many introductions end with a brief summary statement and roadmap indicating the scope or division of the paper that follows.

Italics. While italics were once standard for all foreign words and phrases, today the trend is to drop them in favor of roman (standard) face when the foreign term is well known. Thus most, although not all, publications now use "in utero," "in vivo," "in vitro," "in situ," "i.e.," "e.g.," "ca.," "ad hoc," "vice versa," "a priori," and "status quo," rather than their italicized counterparts.

Use italics for unfamiliar foreign terms:

Identical cell bodies depend, *ceteris paribus*, upon identical nucleoplasm.

Indicate italics in a manuscript by underscoring the term: ceteris paribus.

Use italics for titles of journals, books, and newspapers, including abbreviations: *IEEE Spectrum*.

Use italics for genus and species so long as the genus is singular (see also **Genus names**); for legal citations; and for single letters: All his *p's* looked like *q's;* Planck's constant, *k*.

Use italics to introduce terms, but do so consistently.

Not

Type I Superconductor: a superconductor having only one critical magnetic field, below which currents flowing over the surface cancel a magnetic field inside the material.
Type II Superconductor: a superconductor that has a second, higher critical magnetic field, owing to its ability to trap magnetic flux inside the material.

But

Type I Superconductor: a superconductor having only one critical magnetic field, below which currents flowing over the surface cancel a magnetic field inside the material.

Type II Superconductor: a superconductor that has a second, higher critical magnetic field, owing to its ability to trap magnetic flux inside the material.

For italics in mathematical and statistical expressions, see **Mathematical expressions** and **Statistics, symbols used in.**

J

Jargon. Define unfamiliar technical terms either immediately or in a glossary. Given the increasing specialization of science and technology, specialized terms within any one field are very likely to be unfamiliar to those outside that field, however technical their training. See also **Glossary** and **Definitions of terms.**

Joint possessives. Use the possessive with the last name: *Garetz and Lombardi's results.*

Justified right vs. ragged right. Text that is right-justified is aligned along the right margin. The pages of this book are justified on both sides. To align copy, space is added or deleted.

K

Kelvin. No degree sign is used: 0 K.

Key words. Some publications and companies require a list of 3–8 key words or descriptors to summarize main points in a document. The key words accompany the abstract. Choose these words with care, as they may be the basis for indexing and retrieval.

Boelaert, J., de Jaegere, P. P., Daneels, R., Schurgers, M., Gordts, B. and van Landuyt, H. W. Case report of rental failure during norfloxacin therapy. *Clin. Nephrol.* 25(5): 272, May 1986 (in Letters to the Editor).

Key words: Norfloxacin, desipramine, levomepromazine, melperone, viloxazine, ciprofloxacin, piromidic acid.

Abstract: This is the first report of acute renal failure during norfloxacin therapy. The patient, a 75-year-old woman, had received levomepromazine (5 mg/day) and viloxazine (100 mg/day) for 1 and 2 mo, respectively, for depression. Despiramine (25 mg tid) was started on July 29, 1985, and was increased to 50 mg tid on August 5, 1985, when melperone (100 mg/day) was initiated. Therapy with norfloxacin (400 mg bid) was started on August 1, 1985 because of urinary tract infection due to Escherichia coli. Before treatment, serum creatinine and urea values were normal; however, on the sixth day of therapy these values were 451 mmol/L and 10 mmol/L, respectively. The results of a lymphocyte transformation test with norfloxacin were negative. Histological examination revealed the presence of intense interstitial edema with a discrete multifocal lympho-plasmocytic infiltrate without eosinophils. Norfloxacin and desipramine were discontinued and renal function returned to normal. Rechallenge with desipramine (25 mg tid) did not alter renal function. Although the pathogenetic mechanism of norfloxacin-induced renal failure is not clear, the histological findings suggest a toxic rather than an immunologic mechanism. Because adverse renal effects have also been reported for piromidic acid and ciprofloxacin, the correspondents suggest that the renal safety of quinolone antibiotics should be further examined.

L

Latin terms. See **Greek and Latin words and expressions.**

Leading (rhymes with "sledding"). The amount of white space between the top of one line and the top of the next line. It is measured in points. See also **Points.**

Legends. See **Captions.**

Lists. To itemize a short list of words, phrases, or sentences run into text, use arabic (1, 2, 3) or lowercase roman (i, ii, iii) numerals. Many publications ask that you enclose numerals with parentheses: *(1),* not *1).*

Follow normal punctuation rules for the colon if you use it to introduce the list; that is, use no colon between preposition and object, verb and object, or verb and complement. See **Colon.**

Use initial capitals when each item in a displayed list is a complete sentence. Close each sentence with a period. When displaying phrases, begin each phrase with a lowercase letter; no punctuation is necessary to separate items. Close the list with a period.

For displayed lists (lists in which each item begins on its own line), use arabic numerals followed by periods (1. 2. 3.) or, if order is not important, indentation devices such as bullets (•, •, •) or squares.

Make all items in the list parallel. If, for instance, one item begins with a phrase, so must the others. To display items effectively, separate the list from the introductory text by a carriage return. Use either tabs or hanging paragraphs to make each item visually distinct.

not

Here are other points to bear in mind when proofreading:
1. Check all data against your original manuscript. Of special concern are the numbering of equations and the spelling of foreign names.
2. Check all citations carefully.

but

Here are other points to bear in mind when proofreading:
   1. Check all data against your original manuscript. Of special concern are the numbering of equations and the spelling of foreign names.
   2. Check all captions carefully.

Literature cited. See **References, in text** and **References, lists of.**

Literature reviews. See **Review articles.**

M

Margins. White space surrounding printed page. The inner margin between the text and the binding is the **gutter.** The white space above the top line is the **head.** The other margins are the **outside** and the **foot.**

Mathematical expressions. Use italics for mathematical symbols, including the following:

$\simeq$ approximately equal to
$\propto$ proportional to
$\sqrt{a}$, $a^{1/2}$ square root of a
$\bar{a}$ mean value of a
$\ln x$ natural logarithm of x
$|a|$ absolute value of
$\equiv$ identically equal to
$\sqrt[n]{a}$, $a^{1/n}$ nth root of a
$\log x$ logarithm to the base 10 of x
$\exp x$ exponential of x

Do not use italics for trigonometric functions, abbreviations, numerals, or operators:

cos [but cos(ax)], cot, det, dim, exp, ln, log, max, min, mod, sin, tan

Use boldface rather than italics for arrays, vectors, and tensors and for summation, product, and infinity signs: $\mathbf{A \cdot B}$, $\boldsymbol{\Sigma}$.

Skip a space before and after operators ($=$, $-$, $\times$, $=$).

Break equations before the operator sign. The sign appears at the beginning of the new line:

$$y = v + 2$$
$$= \sqrt[3]{\left[-\frac{1}{2}r + \sqrt{\left(\frac{1}{27}f^3 + \frac{1}{4}r^2\right)} \right]}$$
$$+ \sqrt[3]{\left[-\frac{1}{2}r - \sqrt{\left(\frac{1}{27}f^3 + \frac{1}{4}r^2\right)} \right]}$$

Number displayed equations (equations that are on lines by themselves rather than run into the text) with consecutive arabic numerals, enclosed in parentheses either flush left or flush right. Do not display short equations if they are not referred to in subsequent text.

Omit punctuation after a displayed equation, but retain punctuation before it.

According to Snell's law,

$$n_1 \sin\theta_2 = n_2 \sin\theta_2$$

Separate two displayed equations on the same line by space, not punctuation. If the equations are joined by a conjunction, these, too, are set off by space.

$$r = \sqrt{x^2 + y^2} \qquad \text{and} \qquad \theta = \tan^{-1}\frac{y}{x}$$

If you do not have italic or boldface fonts, indicate italics by underlining and boldface by a wavy underscore. If you do not have Greek fonts, circle Greek letters represented in roman (v, k, u, n for ν, κ, μ, η) on first use and explain that a Greek font is needed. Distinguish between 1 and l, 0 and O, ' and the mark for prime.

Follow the order of parentheses, square brackets, and braces, beginning with parentheses for the innermost quantity.

$$v = \{ -(r/2) + [(q/3)^3 + (r/2)^2]^{1/2}\}^{1/3}$$

Measures, units of. See **SI units.**

Modifiers, misplaced or dangling. Modifiers include adjectives, which modify (explain) nouns or pronouns; and adverbs, which modify verbs, adjectives, and other adverbs. Modifiers may be phrases or clauses. Place modifiers next to the words they modify. For instance, in the sentence *The panel on nutrition said there is considerable evidence to link dietary fats with cancer in its report*, the expression *in its report* is misplaced. It is meant to modify *The panel on nutrition said* and should be placed closer to these words: *The panel on nutrition said in its report that there is considerable evidence to link dietary fats with cancer.*

Readers expect introductory phrases to modify the subject of the main sentence. Introductory phrases that fail to do this are said to dangle. In a sentence like *Currently assigned as a sorter, my duties include processing parcels to their respective districts*, the expression *Currently assigned as a sorter* dangles, as it incorrectly modifies *my duties* instead of modifying the person doing the sorting. *Currently assigned as a sorter, I have many duties, including processing parcels to their respective districts.*

Some sentences with misplaced or dangling modifiers may be corrected by moving the modifying phrase to its proper position. *Like its predecessors, the audience found this report puzzling* could easily be changed to *The audience found this report, like its predecessors, puzzling.* (*Like its predecessors* should modify *report*, not *audience*.) *Once placed in a refereed journal, scientists throughout the discipline will read the findings* can be corrected by moving the modifier *Once placed in a refereed journal* next to the word it modifies, *findings.*

Some sentences are best corrected by adding a logical subject. *To change screen height, the blue button must be depressed* could be revised: *To change screen height, the user must depress the blue button* or *Press the blue button to change screen height* or *To change screen height, press the blue button.*

Some sentences must be recast to avoid ambiguous modification. In *That he finished the job completely amazed the supervisor*, "completely" may refer either to finishing the job or to amazing the supervisor. See also **Only, placement of.**

N

Namely. Like "that is" and "for example," "namely" is considered parenthetical. Set it off with commas.

Negatives. Because they are indirect (*does not have* for *lacks*, *did not remember* for *forgot*), negatives are easily misread. Prefer the positive for commands. See **Instructions and procedures.**

Nomenclature. As disciplines evolve, they develop their own systems for naming, symbolizing, and coding. Responsibility for coordinating practices is often in the hands of a committee of the related professional society; for instance, the International Union of Pure and Applied Chemistry (IUPAC) coordinates all practices for chemical nomenclature. Sometimes nomenclature is collected in a handbook; for instance, pulmonary and respiratory terminology is found in the *Handbook of Physiology: A Critical, Comprehensive Presentation of Physiological Knowledge and Concepts.* Updated terminology often appears regularly in a journal within the field. Virus terminology, for instance, appears in summary reports of the International Committee on Taxonomy of Viruses (ICTV) in the journal *Intervirology. The International Journal of Systematic Bacteriology* publishes guidelines for the names of genera and species. Some standard sources of nomenclature within disciplines follow: **Chemical:** Cahn, R. S., Dermer, O. C. *An Introduction to Chemical Nomenclature,* 5th ed. London: Butterworths, 1979. *IUPAC Nomenclature of Inorganic Chemistry,* 2nd ed. Oxford: Pergamon Press, 1981. *IUPAC Manual of Symbols and Terminology for Physicochemical Quantities and Units.* Whiffen, D. H. Oxford: Pergamon Press, 1979. *IUPAC Nomenclature of Organic Chemistry, Sections A,B,C,D,E,F,H.* Elmsford, N.Y.: Pergamon, 1979. **Physical.** *IUPAP, Symbols, Units and Nomenclature in Physics.* International Union of Pure and Applied Physics. Published in *Physica* 1978, 93A, 1. **Drugs.** *The Merck Index: An Encyclopedia of Chemicals, Drugs, and Biologicals,* 10th ed., Rahway, N.J.: Merck, 1983; Griffiths, M. C., Fleeger, C. A., Miller, L. C. eds. *USAN and the USP Dictionary of Drug Names,* Rockville, Md: U.S. Pharmacopeial Convention Inc., 1985. Fuerst, S., Van Laan, S. eds. **Genetics.** Instructions to authors in *J Biol Chem,* 1987; 262:1–11. **Animal Genetic Terms.** Staats, J. Standardized nomenclature for inbred strains of mice: eighth listing. *Cancer Res.* 1985; 45: 945–977.

Nominalizations. Nominalizations are nouns derived from verbs or adjectives: *analysis* or *make an analysis of* for *analyze, has a corrosive effect on* for *corrodes.* Nominalizations are indirect, verbs direct; verbs state the action, nominalizations say the action exists. Writers who pair nominalizations with passives will find themselves producing wordy, indirect sentences. A clear statement like *The Board recommends we adopt the conclusions* emerges as *A recommendation has been made by the Board that the conclusions be adopted.*

None. See **Number of subject and verb.**

Nonrestrictive vs. restrictive words, phrases, and clauses. Nonrestrictive information is parenthetical. Use "which" to introduce nonrestrictive clauses. *The nomenclature, which we found inconsistent, was completely revised.* (All of the nomenclature was revised.) Restrictive information is essential, rather than parenthetic. Use "that" to introduce restrictive clauses. *The nomenclature that we found inconsistent was completely revised.* (Only the nomenclature that was inconsistent was revised.) Set off nonrestrictive information with commas. Do not set off restrictive information with commas. Not *Women, who are more than thirty-five years old, show reduced fertility.* but *Women who are more than thirty-five years old show reduced fertility.* See also **Comma.**

Not only, but also. See **Parallel construction.**

Number of subject and verb. Subjects and verbs should agree in number: A singular subject requires a singular verb, a plural subject a plural verb. Compound subjects (subjects joined by "and") require a plural verb (*The compilation of information and its graphic representation are both emphasized*) unless they represent a single idea (*The long and short of it is that we are leaving*).

If the compound subject is modified by "each" or "every," the verb is singular. *Each computer and printer is available for the stu-*

dent. Collective nouns are usually construed as singular (*The staff has decided. . . . Added was 50 milliliters*); however, collective nouns may be construed as plural if each member is acting individually: *A number of engineers are switching to management* but *The number of engineers is decreasing.*

"Each," "neither," and "either" are always singular when they are pronoun subjects. (*Neither of the instruments works well.*) Ignore objects of prepositions and compound prepositions such as "along with," "in addition to," "together with," and "accompanied by" when determining the number of the verb. *The head of the lab, as well as her assistant, visits the site each week.*

"None" may be singular or plural, depending on whether you mean "not a single one" (singular) or "all are not" (plural). The trend today is to construe "none" as a plural. If you mean "not a single one," try substituting "not one" or "no one" for "none." *None of the applicants were as skilled as their predecessors.* Noun clauses that are the subject of the sentence take a singular verb. *What the manager decides is his own business.* Linking verbs (*am, is, are, was, were, seem, appear*) take their number from the subject. Thus *Repeated absences were the reason he was fired* but *The reason he was fired was repeated absence.*

Numbers, figures vs. words. Use figures with units of measure (*0.50 mg, 5 sec, $68, 44 mL*); in a mathematical or chemical context (*5 orders of magnitude, factor of 8, the 1s orbital of hydrogen*); for items and sections (*Sample 7, Unit 3*); and for figures and tables (*Fig. 1, Table 3*).

Use words for numbers below ten, both cardinal and ordinal, unaccompanied by a unit of measure: *three pipettes* but *30 pipettes; fifth trial* but *15th sample. Only five people appeared for the committee meeting* but *Only 11 people voted for the motion.*

Use words for numbers below ten, even when accompanied by a unit of measure, if the use is not technical: *This assignment was given to me seven years ago at the UCLA School of Medicine.*

Use words for numbers that begin a sentence. If a unit of measure accompanies the number, spell it out, too. (*Fifteen milliliters of water was added.*) Avoid spelled-out numbers by recasting the sentence. (*Added was 15 mL of water.*)

Use words for common fractions (*one-fourth of the sample; one-half inch; two-tenths; one-thirtieth*).

If numbers in the millions and above occur in narrative text, use a figure followed by the word "million," or "billion": *1.3 million tons, 555 billion gallons, $4.5 million, $5 million to $15 million* (but never *$5 to $15 million,* unless you mean the first item to be five dollars).

If you use two consecutive numerical expressions, either spell out one expression, or recast the sentence: *Twenty-seven 6-packs came with the set.*

In a series where one number might ordinarily be spelled out and another shown in figures, use all figures if the series has one item of 10 or greater: *In the group, 19 people were treated for pansystolic murmur and 7 for atrial systolic murmur . . . 3 pipettes, 5 vials, and 12 flasks* but *Two test tube racks, 17 test tubes, and 2 centrifuges filled the top of the lab bench.* ("Two" begins the sentence and must be written out.)

These rules also apply to adjectival forms: *five-year-old violinist; 30-hour trial; two-photon spectroscopy.* See numbers also in **Copy editing marks, Decimals, Figures, Fractions, Hyphen, Lists, SI units,** and **Tables.**

O

Omission. To show the deletion of a phrase, use a comma. *The alkaline phosphatase was 101 U per liter (normal, 41 to 133).* The use of a comma for deletion is particularly common in the second element of a balanced sentence. (One of her major sources for nomenclature is the *USAN and the USP Dictionary of Drug Names;* the other, *The Merck Index.*) To show an omission in quoted material, use points. See **Ellipses.**)

One of those. In clauses that follow the expression "one of those," use the plural. *This is one of those spectrometers that are always breaking.*

Only, placement of. Place *only,* like *just,* immediately before the word it modifies. Compare the meaning in these sentences based on the placement of *only. Only she saw six patients a day.* (Everyone else in the practice proceeded differently.) *She only saw six patients a day.* (She spoke to the rest on the telephone.) *She saw only six patients a day.* (Her practice is small and exclusive.) The common error is to use *only* too soon in the sentence.

Ordinal numbers. Spell *first* through *ninth* unless they are part of a series with an ordinal greater than nine. (*The third and fifth sample* but *The 3rd, 4th, and 10th sample*). Use words, not figures, for ordinals that begin sentences. (*Fifth in the series was Sample 127.*)

Use figures for ordinals greater than ninth, except at the beginning of a sentence. See also **Numbers, figures vs. words.**

Organizations. Although some organizations and agencies may be known by abbreviation rather than full name (CDC for the Centers for Disease Control, VA for Veterans Administration), use the full name on first mention in text, figure reference, or abstract, followed by the abbreviation enclosed in parentheses. Unexpanded abbreviations pose difficulties for people outside a field. See also **Abbreviations and acronyms.**

Capitalize all major words in the names of organizations and agencies, as well as divisions and departments of organizations. Use lowercase for definite and indefinite articles, conjunctions, and short preposition unless these are the first or last words. Do not capitalize "the" if it is not part of the title: *the Department of Health, the Centers for Disease Control.*

For possessives, use 's only with the last name. *SmithKline Beecham's pharmaceutical products.*

Orientation. A printer's term for the direction of the print on the page. See **Portrait vs. landscape.**

Orphan. One short word (sometimes two) left alone on a line at the end of a paragraph.

Overlay. Transparent paper placed over artwork to protect it, or to show instructions for cropping, arrows, or other notation that cannot be placed directly on the photograph or artboard.

Over vs. greater than. See **Greater than or more than vs. over.**

P

Page proof. See **Proof.**

Paragraph. Paragraphs are distinguished by the properties of *unity* and *continuity. Unity* means that all the sentences in the paragraph are about the same topic or idea. *Continuity* means that the sentences expressing or developing the idea flow logically from one to the next. This flow is often promoted by the use of transitions or repetition of key words. See **Transitions.**

Page-length paragraphs, common in nineteenth-century writing, are rare today, but no absolutes exist for paragraph length. One-sentence paragraphs, once frowned upon, are now acceptable if dictated by content. Editors tend to shorten paragraphs dramatically when columns are very narrow to avoid a wall-of-print effect.

Parallel construction. In parallel construction, all the items joined in a series or comparison have the same grammatical form: adjectives are linked with adjectives, prepositional phrases with prepositional phrases, infinitives with infinitives. For example, in William Harvey's statement that "the organ is seen now to move, now to be at rest; there is a time when it moves, and a time when it is motionless," the contrasting phrases "to move" and "to be at rest" are parallel because they are both infinitives. "When it moves" and "when it is motionless" are parallel clauses linked by "and."

The writer should keep the parallel when using the correlative conjunctions *either/or, neither/nor, not also/but also, both/and.* Avoid *"He is not only a scientist, but also he practices law."* (The words following "not only" are "a scientist." The words following "but also" are "he practices law." "A scientist" and "he practices law" are not parallel.) Try "He is not only a scientist, but also a lawyer" or "He is not only a scientist, but a lawyer as well."

Avoid errors in suspended parallels. Not *Her work was, and continues to, be quite promising* but *Her work was, and continues to be, quite promising.*

When listing items preceded by definite or indefinite articles, keep the list parallel by using "the" or "a/an" either once at the beginning or after each item. Either *The SI is based on seven fundamental units: the meter, kilogram, second, mole, ampere, degree kelvin, and candela.* Or *The SI is based on seven fundamental units: the meter, the kilogram, the second, the mole, the ampere, the degree kelvin, and the candela.* But not *The SI is based on seven fundamental units: the meter, kilogram, second, the mole, ampere, degree kelvin, and the candela.*

Keep the parallel in split figure captions: Not *Fig. 1. (a) opening cavity, (b) to close cavity* but *Fig. 1. (a) opening cavity, (b) closing cavity.*

Keep the parallel in lists.

Not

Submitting a Manuscript
1. Double-space all copy.
2. Number all figures.
3. You should prepare a separate legend manuscript.

but

Submitting a Manuscript
1. Double-space all copy
2. Number all figures
3. Prepare a separate legend manuscript.

Not

Making an Index
   1. *Selecting Entries.* Select and phrase entries from the reader's point of view.
   2. *Annotating Proof.* On duplicate galleys, mark all words or phrases to be used as main entries. Use double underlining for subentries.
   3. Transfer entries to 3 × 5 cards. List each main entry on a separate card. Repeating the main entry at the top, list each subentry on a separate card. Include the galley or page number for each main entry and each subentry.

(The list is not parallel because the first two items start with a phrase, but the third item begins with a complete sentence.)
but

Making an Index
   1. *Selecting Entries.* Select and phrase entries from the reader's point of view.
   2. *Annotating Proof.* On duplicate galleys, mark all words or phrases to be used as main entries. Use double underlining for subentries.
   3. *Preparing the Cards.* Transfer entries to 3 × 5 cards. List each main entry on a separate card. Repeating the main entry at the top, list each subentry on a separate card. Include the galley or page number for each main entry and each subentry.

Parentheses and brackets. Use parentheses for figure and table citations: *The computed tomographic scan of the cranium (Fig. 2) showed evidence of a recent Caldwell-Luc operation.*

Use parentheses to introduce abbreviations, acronyms, and definitions: *A computerized tomographic (CT) scan of the cranium showed evidence of a recent Caldwell-Luc operation. The Health Care Financing Administration has contracts with 54 peer review organizations (PROs) to monitor hospital use and quality of care for Medicare patients. Acetylene (C_2H_2) and ethane (C_2H_6) have been identified on Neptune.*

Use parentheses for interpolations and asides: *In addition, PROs could not be affiliated with a health care facility or association of facilities or with a third-party payer. (The Idaho PRO, awarded to the local Blue Cross, is the only exception.)*

Use parentheses for subsidiary information that clarifies procedures or results: *The alkaline phosphatase was 101 U per liter (normal, 41 to 133). The urea nitrogen was 8 mg per 100 mL (3 mmol per liter). Virtually half the respondents (13 of 27) accepted the false suggestion as real. Patients could be identified who hyperabsorbed calcium and had a coincident marked reduction in the parathyroid hormone-1,25-$(OH)_2D$ axis (e.g., Patients 1 and 2, Table 2), whereas in others who hyperabsorbed calcium, the axis was essentially nonsuppressible (e.g., Patients 3 and 4, Table 2).*

Use parentheses to direct the reader's attention in figure captions.

Fig. 2. Axial CT Sections. The scan at the level of the maxillary antra (top) reveals absence of the anterior half of the nasal septum. . . . The scan at the level of the midorbit (bottom) reveals a surgical cavity in the right ethmoid sinus. . . .

Use parentheses to enclose numbers of items in lists that are run into text, rather than displayed. *More research is needed on the impact of the changes (1) on different patient populations, (2) on hospitals and their employees, (3) on the neighborhoods that surround them.* See **Lists.**

Use parentheses to identify brand names and manufacturers, particularly in the procedural section of a paper. *Recommended was captopril (Capotan), in a maximum maintenance dosage of 100 mg tid.*

Use parentheses for citations:

A report of intramuscular lidocaine for prevention of lethal arrhythmias in the prehospitalization phase of acute myocardial infarction appears this month in *The New England Journal of Medicine* (1985; 313:1105–1110).

Do not use parentheses for figures or references when the information is essential, rather than parenthetical. Not *In Fig. (1), the cavity of the right ethmoid sinus is shown* but *In Fig. 1, the cavity of the right ethmoid sinus is shown.*

Use no period for parenthetical sentences within sentences: *The cavity of the right ethmoid sinus (see Figure 1) shows severe damage.* If the parenthetical sentence is outside the sentence it clarifies, use a period: *The cavity of the ethmoid sinus shows severe damage. (See Figure 1.)*

If the text before the parentheses calls for a comma, put it after the closing parenthesis mark. Do not use a comma before a parenthesis mark. Not *According to Smith, (currently on leave from MIT) the findings are mixed* but *According to Smith (currently on leave from MIT), the findings are mixed.* Do not use dashes with parentheses.

Use either parentheses or brackets, depending on the guidelines of the publication, to enclose reference citations. Commas and periods that are part of the text precede citation numbers; semicolons and colons that are part of the text go after the citation numbers.

The hypothesis that as much as 80 percent of cancer could be due to environmental factors was based on geographic differences in cancer rates and studies of migrants. (136)

Several reports describe the value of magnetic resonance imaging (MRI) in visualizing the lesions of multiple sclerosis. [1, 3, 5]

Use brackets in quotations to show that the material within the brackets was not interpolated by the source quoted:

Feynman comments that "the phenomenon of colors produced by the *partial* reflection of white light by two surfaces is called iridescence [italics added]."

Use brackets for parentheses within parentheses. For chemical and mathematical notation, use parentheses for the innermost item, followed by square brackets and braces.

$$v = \{-(r/2) + [(q/3)^3 + (r/2)2]^{1/2}\}^{1/3}$$

Use brackets to indicate concentrations: $[NaCl] = 2.25 \times 10^{-4}M$

Passive voice. See **Active vs. passive voice.**

Percent. In general, use the symbol with figures when giving results: *44% of the response.* Include the symbol for each number within a range including zero: *from 0% to 15%.*

Use the word "percent" for approximate or rounded numbers. *At least 10 to 20 percent of the mast cells appeared degranulated in the index case.* (Also widely seen, *per cent,* although *percent* is gradually replacing it.)

If you are discussing a percent derived from a small sample, give the numbers from which the percentage derives. *Within the group, 48% (13 of 27) accepted the false suggestion as real.*

Period. Place periods, like commas, before closing quotation marks or citation numbers. *According to Dr. Sinclair, "a convention has been established to include hyperkalemia as a primary term." Used was a one-sided sequential testing procedure.*[1]

Use periods after the indented arabic numbers that itemize a displayed list. See **Lists.**

Use periods to represent decimals. See **Decimals.**

Do not use periods after units of measure, scientific symbols, or technical abbreviations unless the abbreviations might be misread as words. (Thus *no., tan., at. weight.* Some exceptions: *a.m., ca., ed., e.g., ibid., i.d., i.e., p.m., v., vs.*) See **Abbreviations and acronyms.**

Do not close titles of papers or reports with periods.

See also **Ellipses.**

Persuade. See **Convince, persuade.**

pH. The pH scale is one of the non-SI systems retained for reporting. (See **SI units.**) Avoid beginning sentences with abbreviations, but if pH must be used to begin a sentence, do no capitalize the "p." Do not capitalize the "p" in titles. Do not use italics.

Photographs. See **Figures.**

Pica. The pica is a printer's unit of measure: Six picas occupy about one inch or 12 points. On the typewriter, pica is a size of type that fits 10 characters per inch (cpi); elite fits 12 cpi. See also **Point and pitch.**

Pitch. See **Point and pitch.**

Plurals. Do not add *s* to plural units of measure: *3 mL* not *3 mLs.* Form the plural of solid compounds by adding an *s* at the end of the word: *cupfuls.* For plurals of all-capitalized abbreviations (ICBM), add *s* without an apostrophe: *ICBMs.* For plurals of numbers, add *s* without an apostrophe: *0s, 1920s.* For plurals of symbols, signs, and letters, use an apostrophe: *θ's, p's* but *Ss* for *subjects.* Use italics for words, letters, and numbers, but not for signs and symbols. See **Abbreviations and acronyms** and **Apostrophe.** For plurals of Greek and Latin terms, the tendency is to use the anglicized (English language) version for many, although not all, terms. See **Greek and Latin words and expressions, Number of subject and verb,** and **Collective nouns.**

P.M. See **A.M., P.M.**

Point and pitch. Fonts may be characterized by both pitch (characters per inch) and point (the height of the characters). This is an example of 10-point 12 pitch-type. This is an example of 07-point 16.66 pitch. The pitch or number of characters per inch depends on whether the type is proportional or fixed (nonproportional). In fixed pitch, all the characters occupy equal space. The "w," for instance, occupies the same space as the "i." This is an example of 10-pt. fixed pitch. Fixed pitch-fonts are best for aligning columns. In proportional pitch, spacing varies with each character. The "t," for instance, takes less space than the "w." This is an example of 10-pt. proportional pitch. See also **Font** and **Typeface.**

Points. Points are a printer's measure for the height of characters. There are 72 points to the inch. Books and reports are commonly printed in 10-, 12-, or even 14-point type for text, and slightly smaller type for footnotes. The amount of white space between the top of one line and the top of the next line (called "leading") is also measured in points. Here are examples of point size:

Aa	Aa	Aa	Aa
(8 pts)	(10 pts)	(12 pts)	(14 pts)

Portrait vs. landscape. Portrait mode is material printed across the width of the page; landscape mode is material printed across the length of the page. Tables and graphs that can be viewed upright are in portrait mode; tables and graphs that must be viewed sideways are in landscape mode (also called side turns). Try to design tables and graphs so that they can be viewed upright.

Possessives. In general, form the possessive singular by adding "'s" to the noun, including nouns that end in "s." Pronominal possessives (*its yours, his, hers, ours, theirs*) require no apostrophe. For multiple owners, use the possessive form only with the last owner: *Rushdale and Clement's results*. The possessive is rarely used with two-part eponyms. See **Apostrophe** and **Eponyms.**

Prefixes. See **Hyphen** and **SI units.**

Principal vs. principle. A principle is a guiding rule: *Graduate students learn the principles of scientific investigation.* "Principal" is an adjective meaning first in importance: *The principal problem with any liquid crystal display is its readability.* "Principal" as a noun refers to a head or central performer. *Dr. T. Kilson is the principal in the investigation.*

Prior to. Use "before."

Procedural (methods or experimental) section. In the procedural section of a scientific or technical report, give sufficient detail so that other qualified workers could reproduce the steps described. Standard procedures do not need to be recounted if the details are published elsewhere. Instead, give a reference: *The data were collected as part of a case-control surveillance system.*[10]

Prefer generic terms but use brand names if the information is necessary for others to understand and reproduce the experiment. See **Brand names and trademarks.**

Use a separate paragraph labeled "Cautions" to describe any hazardous procedures. See also **Instructions, SOPs (Standard Operating Procedures), and procedures.**

Progress reports. Traditionally short and to the point, these reports are prepared at intervals (monthly, quarterly, semiannually, annually) to summarize developments either on a particular project or on overall accomplishments during the interval. For instance, *monthly reports* may begin with a summary of the month's work and then comment on the upcoming work in relation to possible problems and proposed solutions. *Quarterly reports* may begin with a summary of the work done in the past three months. A formal section with results and interpretation may follow. Results may be supplemented by diagrams, tables, or sketches. Usually the writer discusses future actions in relation to the timeframe and ultimate objectives. See also **Reports technical and scientific.**

Pronouns. Pronouns must have a noun or pronoun as the antecedent, not an adjective. Not *The tenacity of the American effort suggested their certainty of success.* (The antecedent of "their" is "American." Change antecedent to "Americans.")

Antecedents should be clear. If the reader must pause to decide on the probable antecedent, recast the sentence to prevent any momentary jolt, ambiguity, or misunderstanding. Not *NASA engineers, acting on the prodding of the special investigatory committee, presented a series of design changes for the O-rings, which pleased the committee.* (The antecedent of "which" may be either design changes or the fact that the engineers responded based on prodding.)

Pronouns should agree in number with the words they replace: singular antecedent, singular pronoun; plural antecedent, plural pronoun. Not *Each animal in the test group was isolated for six weeks before their examination* but *Each animal in the test group was isolated for six weeks before its examination.*

The indefinite pronouns "each," "either," "everyone," "everybody," "anybody," "anyone," "either," "neither," "one," "no one," "someone," and "somebody" are singular. "Several, "few," "both," and "many" are construed as plural. "All," "any," "some," "most," and "none" are singular or plural depending on meaning or referent. *Some of the applicants are submitting their papers. Some of the carbon silicate was emptied from its container.*

Many mistakes come about when the antecedent is "everyone," "anybody," "each," or "either." These antecedents take a singular pronoun: *Each of the men in the research group had his own way of looking at the problem.* When women enter the research group, the problem becomes thorny, as English possesses no singular pronoun that includes both sexes. Historically, the solution has been to use "him" or "his" to refer to singular antecedents: *Each staff member should do as he pleases.* The conscientious writer who wants to include females often tries, "Each staff member should do as they please." This creates a new problem. "They" is incorrect, for it is plural, and does not agree with the singular antecedent, "each staff member." One solution is to use a plural: *Staff members should do as they please.* When the plural won't work, the resourceful writer must try either recasting the sentences or alternating male and female examples. See also **Gender.**

Antecedents of collective nouns take a singular or plural pronoun depending on meaning. Compare *The staff submitted its demands.* (The staff acting as one unit.) *The staff are undecided about their demands.* (The staff acting as individuals.) See also **Collective nouns.**

Use reflexive pronouns when the subject is acted upon (*To test his reflexes, he hit himself on the patella*) or when you want to use rhetorical emphasis (*I myself am opposed to this change in research funding*). Do not use reflexive pronouns as subjects or objects to avoid using "I/we" or "me/us." Not *Sinclair, Lewis, and myself visited the installation* but *Sinclair, Lewis, and I visited the installation.* Not *Give the evaluations to Sinclair, Lewis, or myself* but *Give the evaluations to Sinclair, Lewis, or me.*

Use "which" for nonrestrictive clauses, "that" for restrictive clauses. Not *The term "serum" precedes the element which is being tested* but *The term "serum" precedes the element that is being tested.* See **Nonrestrictive vs. restrictive words, phrases, and clauses** and **That vs. which.**

Do not use " 's" with possessive pronouns (*yours, its, his, hers, ours, theirs, whose*).

Proof. Printed or hard copy of typeset materials. Proof that is not yet in page form is **galley proof** or **galleys.** Proof made into page form is **page proof.** Proofreading marks for galleys or page proofs are put in the margin, not above the line.

Proofreading marks. Marks used to indicate errors on proofs. All proofreading marks go in the margins of the typeset proofs. Copy editing marks, in contrast, usually go on or above the line.

insert ∧
delete ℓ

delete and close
add space
delete space
move left
move right
lower
raise
center
italics
boldface
boldface italics
flush left
flush right
roman
superscript
subscript
apostrophe
quotation marks
comma
period
semicolon
colon
lowercase
lowercase with initial
   capitals
capitals
let text stand as set
indent 1 em
indent 2 em
indent 1 en
paragraph
run in (go to next
   page or paragraph)
en dash
em dash
hyphen
equals sign
prime sign
parentheses
brackets
wrong font
transpose
approximately
align vertically
align horizontally

Proposal. A proposal (from *pro*, meaning "forth," and *poser*, "to place or put") is an offering that the writer puts forward, typically to solicit new business, to compete for research money, to bid for a government contract, or to justify a new expenditure for equipment. Proposals may be unsolicited (written by authors without a request from a funding source) or solicited [written to answer a request for a proposal (RFP) from funding sources such as the Environmental Protection Agency].

By their nature, proposals must be persuasive documents, telling not only what the author wants to do, but why the author is the best person to do it, the person with the solution that most clearly benefits the reader. To do this effectively, a good proposal presents a carefully structured plan for how the writer will bring about the proposed solution. The details of the plan may be technical, but the arguments themselves must be persuasive ones that sell the proposer, the proposer's idea, and the competence of the proposer to do the job.

The introduction should establish the need for the writer's solution. If the proposal is written in response to guidelines, the proposer should relate project goals, both short- and long-term, to the stated goals of the funding source. The introduction should also include a statement of benefits. What gains will the project bring? Why do these gains matter? Gains should be stated from the reader's point of view.

Longer proposals may be divided into a technical section, a management section, and a budget. The technical plan should describe what will be done, how it will be done, and when it will be done, including anticipated outcomes. For research proposals, authors should relate the offer to the past and present research that provides its theoretical or methodological framework. Any lacks in this framework that the proposal will fill should be identified. Any ongoing or recently completed work that is conceptually or empirically linked to the proposed work should be discussed. This places the work in a scholarly context and shows how it will help to build the foundation of knowledge within the discipline. For business proposals, writers should show all benefits, for example, reduced cost, improvement in moral, or prestige to the institution.

The management plan should include a brief, salient argument for why the proposer is the right person for the job. Resumes should be in the appendixes, not in the body of the text. A discussion of facilities and equipment should be included if relevant. A prescribed form is usually supplied for the budget. If not, authors should use both figures and an accompanying narrative to explain each line of the budget.

The conclusion should be brief but upbeat, with a restatement of the significance of the problem, the proposed benefit, and any implications that may arise from the work.

Proprietary names. See **Brand names and trademarks.**

Punctuation. See **Apostrophe, Colon, Comma, Dash, Ellipses, Hyphen, Parentheses and brackets, Period, Question mark, Quotation marks,** and **Semicolon.**

Q

Qualifiers. Qualifiers are words that limit the meaning of other words. Compare the unqualified *Everyone responded to the survey* with the qualified *Virtually everyone responded to the survey*. Compare the unqualified *This study shows that hypnosis is not effective to enhance recall* with the more qualified statement, *This study casts doubt on the use of hypnosis to enhance recall*. Common qualifiers include *virtually, it would appear that, it seems that, apparently, suggests that, seems, seemingly, in some ways, in some respects, could, might, rather, little, fairly, quite, usually, often,* and *sometimes*. Also known as "hedges" or even "weasel words," qualifiers are properly used to set limits or restrict meaning. Not *Half the respondents (13 of 27) accepted the false suggestion as real* but *Virtually half the respondents (13 of 27) accepted the false suggestion as real.*

Quasi. Hyphenate when used with an adjective. *Quasi-intellectual, quasi-scientific.*

Query. Question addressed to author or editor on manuscript or proof.

Question mark. No question mark is necessary for indirect questions or for requests: *He asked where the money would come from.*

Quotation marks and quotations. Periods and commas precede closing quotation marks: *Newton wrote that "there is a power of gravity tending to all bodies, proportional to the several quantities of matter which they contain."* Semicolons and colons follow closing quotation marks. *We are aware of the policy stated in the guidelines that "specific test names be recorded on reports strictly within the selection* *provided in the validation file"; our query, however, deals with an entirely different issue—the criteria for the validation file itself.*

Questions marks and exclamation points go either inside or outside of closing quotation marks, depending on meaning. If the quoted material is a question or exclamation, enclose the question mark or exclamation point: *All laboratories using this equipment must post large warnings that include the word "Danger!"* If the quoted material is not part of the question or exclamation, do not enclose the question mark or exclamation point: *Have you filed the "Request for Proposal"?*

Avoid the use of quotation marks to suggest irony. Avoid the use of quotation marks to apologize for or ameliorate the effect of the words used. Not *We would like to clarify some of the "rules" recently issued.* Not *That request really "caught us with our pants down."* Expressions that require apology are best omitted.

Use quotation marks sparingly to indicate the use of a word in an unexpected sense within a technical argument: *A "Christmas tree" is an oil rig that . . . .*

Use quotation marks to enclose words discussed as linguistic units: *The authors decided to define the terms "life span" and "life expectancy," as the two terms are frequently confused.*

Use quotation marks for titles of essays, dissertations, and theses and for names of sections of reports ("Results") and book chapters:

For a more detailed account, please see the "Methods and Materials" section of my earlier paper, *Antibronchoconstrictor Activity of the Intracellular Calcium Antagonist HA 1004 in Guinea Pigs.*

A chapter concerning life span and life expectancy, "Interventions to Increase Longevity," appears in the 1990 edition of *Advances in Medicine.*

Use single quotation marks for quotes within quotes.

Use single quotation marks for quotes within titles or subtitles of papers.

No quotation marks are needed for indirect quotes. *He asked what we would prefer.*

No quotation marks are needed in Question/Answer format, or in discussions that identify each speaker before the quotation.

Keith Holmes: Several questions have been addressed to the panel.

Jane Sinclair: I would like to answer the questions on life span.

Omit quotation marks in block quotes. Set off any quotes within block quotes with double quotation marks.

Punctuation is not always necessary to set off a quotation. Not *Newton said, that "all bodies gravitate toward every planet"* but *Newton said that "all bodies gravitate toward every planet."* When punctuation is necessary, introduce any phrase or single-sentence quotation run into text with a comma. *According to Newton, "there is a power of gravity tending to all bodies" that is proportional to the several quantities of matter which they contain.* Introduce longer and displayed quotations with a colon. *Charles Darwin, always uncomfortable when he had to give a speech, commented later on his sensations during a talk at the Geological Society: "I could somehow see nothing all around me but the paper, and I felt as if my body was gone, and only my head left."* See also *sic* for errors in quoted materials, **Brackets** for punctuation of interpolated elements in quotations, and **Ellipses** for omitted elements.

R

Readability indexes. Readability indexes or formulas use measures like number of syllables in words and number of words in sentences to estimate the reading difficulty of text. Robert Gunning's Fog Index, for example, estimates reading level by taking a 100- to 150-word sample. The user (1) divides number of words by number of sentences. This yields average number of words per sentence. Next, the user (2) counts words of three or more syllables, except for proper nouns, and divides this number by the total number of words in the sample. This yields the percentage of difficult words in the sample. Then the user (3) adds the average number of words in the sentence to the percentage of difficult words, and (4) multiplies the total by 0.4. This yields a Fog count. A count of 10 means the passage should be easy reading for the average tenth grader.

Recto. Right-hand page. The copy editor's instruction "Start recto" means to begin the preface or index, for instance, on the right side.

The **verso** (left-hand page) is the back of any right-hand page. The copy editor's direction "Start recto/verso blank" means the back of the page is left empty.

References, in text (Citations). Cite all references used in the text, as well as in tables, figures, and captions. Acknowledge sources using either an *author and date system* or *a numbering system.*

If you are using an *author and date system,* give author's name and publication date in the text within parentheses (*Mitcham 1988*). Give page numbers or other divisions such as section numbers after the date (*Mitcham 1988, 121–125*). If the author's name appears in the sentence, give the year in parentheses. *According to Morrison (1987), SEED is used as an optical switch.* For two authors, list both (*Thomas & Thomas 1986*); for three authors, list three initially (*Thomas, Thomas, & Morrison 1986*) and then abbreviate (*Thomas et al. 1986*). Place page numbers directly after quoted material. *Weitzenbaum (1986) comments that "the computer is a solution in search of a problem" (p. 122).* If the author has more than one work in the same year listed in the references, use *a,b* to distinguish items. *Fitter (1986a) comments on a weakness in the study.* If no author is available, use the name of the sponsoring agency, institution, corporation, or group. If you cite a source that will not be in your references, given the details within parentheses: *Mr. Sinclair (telephone interview 1 August 1986).*

If you are using a *numbering system,* cite all references used in the text, the tables, the figures, and the captions. Order the references either alphabetically by author or as cited (1, 2, 3). Use arabic numerals either on the line *[Neuroendocrine (3), neuropharmacological (4, 5), and behavioral studies (6–8) suggest that aupaminergic systems modulate endogenous opoid system activity]* or in superscript [Neuroendocrine,[3] neuropharmacological,[4,5] and behavioral studies[6–8] suggest that aupaminergic systems modulate endogenous opioid system activity]. Place semicolons and colons after superscript citations, periods and commas before superscript. If you cite the same material again, avoid repetition by using the number of the original reference. *Most hepatitis infections in neonates are asymptomatic, although fulminant cases are seen occasionally.*[4,20]

References, lists of (Endnotes, Literature Cited, References, Sources). Organize the list either by number or alphabetically by author.

If the list includes references from journal articles, include the author's name and initials, the title and subtitle of the article, the title of the journal (usually abbreviated and italicized), the year of publication, the volume, and the issue month or number when volume pages are not consecutive. Give the complete page span (*70–78,* not *70–8, 108–115,* not *108–15*). Many publications do not require the title of the journal article.

If the list includes books, give the author's name and initials, the editor's name and initials, the italicized title and subtitle of book, the number of the edition except first, the publisher, and the city and year of publication. Include page numbers for specific citations.

Indicate italics by underscoring; indicate boldface with a wavy line. J. Chem. Phys. 45.

Samples follow. Significant variations exist among publications.

REFERENCES

[1] J. D. SIMMONS and S. G. TILFORD, *J. Chem. Phys.* **45**, 2965 (1966).
[2] W. M. McCLAIN and R. A. HARRIS, in *Excited States,* Vol. 3, pp. 1–55 (edited by E. C. LIM). Academic Press, New York (1978).

References

[1]G. Herzberg, *Spectra of Diatomic Molecules* (Van Nostrand Reinhold, New York, 1950).
[2]W. M. McClain, J. Chem. Phys. **55**, 2789 (1971).
[3]W. M. McClain and R. A. Harris, in *Excited States, Vol. 3,* edited by E. C. Lim (Academic, New York, 1978), pp. 1–55.

REFERENCES

1. Kertes P, Hund D. Prophylaxis of primary ventricular fibrillation in acute myocardial infarction: the case against lignocaine. Br Heart J 1984; 52:241–247.
2. Pantridge JF, Webb SW, Adgey AAJ, Geddes JS. The first hour after onset of acute myocardial infarction. In: Yu PN, Goodwin JF, eds. Progress in cardiology. Vol 3. Philadelphia: Lea & Febiger, 1974: 173–178.

Reports, technical and scientific. The term "report" is derived from words meaning "to bear or bring back." A technical report is a verifiable account of events observed by the writer. The style is unadorned, unemotional, and economical. Careful proportion, thorough analysis, and economical presentation are guiding principles.

The format originated in antiquity, probably for forensic investigation. Greek reports were argued in the structure of an introduction; an exposition (the circumstances defining the issue); a proposition that the remainder of the report established as true, valid, or probable; a division or outline of points to be discussed; a proof; a refutation of anticipated objections; and a summary. Revived in the West during the Renaissance, this structure developed in antiquity fits closely the present format for a scientific or technical report. The sequence has been adapted for Ph.D. dissertations, for research published in scholarly journals, for corporate and business reports, and for technical reports written for government contracts or distribution.

The focus of a report depends on its purpose (e.g., to present research results, to review the literature of a field) and its audience. The audience or readership for reports varies from the band of technical experts or peers who read scholarly manuscripts in *Physical Review Letters* to the diversified public readership of *The Surgeon General's Report on Nutrition and Health*. Level of technical detail, use of definitions, and complexity of illustration should be adjusted to allow for both the readers' technical backgrounds and their purpose in reading the document.

Within companies, reports are written at virtually every level to document work. Among the many types of corporate reports are service reports (e.g., inspecting a transformer, analyzing a tax problem, troubleshooting a production snag), research reports (often proprietary), literature reviews, trip reports, topical reports, and progress reports.

Whether a scholarly manuscript or a corporate report, the document must establish main points before introducing detail. To do this effectively, start with an abstract or summary that gives the objective, the findings, and the implications. Use the introduction to sum up the problem and its significance, and, when applicable, to show how the present work adds to or differs from related work on the problem. If the report is very short, substitute a statement of object, scope, and limits for the introduction.

Divide the body of the report logically, using headings and subheadings to show the reader the path of the argument. Define key terms. Use examples and comparisons to bring abstractions to life. Keep the body of the report as lean as possible, using tables for vivid, concise display of data, and graphs for trends. (In-house reports can be shortened by placing supporting materials such as spectra and chromatograms in the appendixes.) Take advantage of headings, tabular layout, boldface and italic type, and displayed lists to divide text and distinguish items.

For readability, try to keep sentences short and language direct. Use active verbs and first person when appropriate.

Be aware of logical divisions in the material, and keep data within the appropriate category. Mixing sections of reports create both logical and organizational problems for readers. See also **Scientific papers.** For parts of the reports, see **Abstracts, Introduction, Key words, Procedural (methods or experimental) section, Results and discussion, Titles.**

Respectively. "Respectively" means "in the order given," not "each" or "one by one." Use *respectively* to link two groups within a sentence. Not *The two lasers have wavelengths of 488 nm and 647 nm respectively* (delete "respectively") but *The patient, a 75-year-old woman, received levomepromazine (5 mg/day) and viloxazine (100 mg/day) for 1 and 2 months, respectively. During the restricted-calcium diet, elevated values for serum immunoreactive parathyroid hormone, nephrogenous cyclic AMP, and 1,25-(OH)$_2$D were observed in 17, 17, and 12 patients, respectively.* "Respectively" is considered parenthetic, and is therefore set off by commas.

Restrictive clauses. See **Nonrestrictive vs. restrictive words, phrases, and clauses.**

Results and discussion. These two sections of a scientific or technical report may be presented separately or combined, depending on the type of data. Results should include a summary of the data and, when appropriate, a discussion of the statistical method. Cite any references for the statistical methods on first mention. Also identify and describe any computational methods employed, including computerized analyses. Include sufficient results to justify the conclusions. Use tables and graphs to display data vividly and concisely, but be sure that tables supplement rather than duplicate the text. Check to ensure that the tables and figures are consistent with information in the text. In the discussion, relate the findings to the original objective. Interpret and compare; point out special features; discuss limitations and future applications of the work.

Use a separate conclusions section if warranted, but not if it simply duplicates information in the discussion.

Review articles. Review articles collect the literature on a subject, summarizing and commenting upon the state of knowledge. The scope may be comprehensive or selective. Review articles may appear in research publications, or they may appear in journals devoted solely to reviews. Some journals devote one issue a year to a review of the literature within the field. Book-length review articles are called "monographs." Many literature searches begin with review articles not only for the overview that they provide, but for the list of primary sources.

River. White space within the printed page that inadvertently creates a pattern of a rivulet or stream, fragmenting the even appearance of the page.

Roman. Regular type, as opposed to italic.

Run back (rb). To shift material from the end of one line to the end of the line above.

Run down (rd). To shift material from the end of one line to the beginning of the next line.

Run in. To merge one paragraph to the next.

Running feet and heads. Identification (e.g., name of journal, title of paper, author's name) that appears at either the bottom or top of the page.

S

Sans serif. Unadorned type style that dispenses with lines projecting from the tops or bottoms of letters. Helvetica is an example of a sans serif typeface.

Scientific papers. The scientific paper emerged in 1665, with the publication of *The Philosophical Transactions of the Royal Society of London* and the *Journal des Scavans* (later the *Journal des Savants*), both published in 1665, followed shortly by the *Acta Eruditorum*, published in Germany in 1682. Since then, the number of scientific journals has multiplied at an astonishing rate: more than 60,000 periodicals were published between 1900 and 1960 alone.

Journal articles, whether theoretical analyses or reports of laboratory experiments, support, extend, challenge, or refute established theory. The work should be both original and significant.

The introduction establishes the research problem, its significance, and the relationship of the author's work to others who have studied the problem. The procedural or experimental section states how the problem was solved in sufficient detail for other qualified scientists to reproduce the steps. The results section, which may be combined with the discussion, summarizes the data and, when appropriate, the statistical interpretation. Results may be displayed vividly and concisely in tables and graphs. Sufficient data should be included to justify conclusions.

In the discussion the author relates the findings to the original objective, interprets, draws comparisons, and points out limitations. Further implications may be discussed in a separate "Conclusions" section if warranted by the content. See also **Reports, technical and scientific.**

Secondary sources vs. primary sources. Secondary sources summarize and sometimes comment upon primary sources. For instance, a literature review or monograph that collects and discusses the findings of the past year on advances in two-photon absorption spectroscopy is a secondary source.

Primary sources included refereed research journals, patents, and dissertations. While journal articles usually provide encapsulated reviews of relevant literature in the introduction to establish the problem, the main function of a research report is to present data that are both new and significant. Therefore while the research report may have useful information on secondary sources, it is a primary source.

Semiannually. "Semiannually" means two times yearly. See **Bi- vs. semi.**

Semicolon. Use a semicolon to link independent clauses. The semicolon provides a closer link than that provided by a coordinating conjunction (*and, but, for, or, nor, so, yet*).

Laser chemistry interests some of the graduate students; others are attracted to rapidly developing areas of polymer chemistry.

Use a semicolon to separate items in a list when the items themselves are punctuated, including names in bylines.

Ruth Sinclair, M.D., Ph.D.; George Henderson, Sc.D.; Henry Eng, B.A.

Susan Kincaid, Department of Chemistry, George R. Harrison Spectroscopy Laboratory, Massachusetts Institute of Technology, Cambridge, MA 02139; and Henry Liu, Department of Chemistry, Polytechnic University, Brooklyn, NY 11201.

The rank order of activity from the most to the least active agent was as follows: ceftizoxime, MIC90 = 0.015 µg/mL; ceftriaxone, MIC90 = 0.06 µg/mL; norfloxacin, MIC90 = 0.125 µg/mL; imipenem, MIC90 = 0.5 µg/mL.

Use a semicolon, not a comma, to link independent clauses joined by a conjunctive adverb (*however, therefore, thus, then, still, hence, indeed, instead, nonetheless, otherwise*). Not *Application of topical, water-soluble antibiotics controls infection in an open wound, however, it also appears to increase inflammation* but *Application of topical, water-soluble antibiotics controls infection in an open wound; however, it also appears to increase inflammation.*

Do not use a semicolon to link a subordinate clause with an independent clause. Not *Arcs are added in an entirely different fashion than nodes; although the accompanying manual fails to clarify the distinction* but *Arcs are added in an entirely different fashion than nodes, although the accompanying manual fails to clarify the distinction.*

Place semicolons after closing quotation marks and after reference numbers.

We have investigated the production of these metastable excited states[13,14]; we are also investigating analogous identity-forbidden transitions in three-photon absorption.[15]

Semimonthly. "Semimonthly" means twice a month. "Bimonthly" means every two months. Try either "twice a month" or "every

two months" to avoid mix-ups. See also **Bi- vs. semi-.**

Shift in person. Do not shift person in the same sequence. Not *Depress the lever and the user raises the screen* (switch from second person "you" to third person "the user") but *Depress the lever and raise the screen* or *The user depresses the lever and then raises the screen.*

sic. Use *sic* sparingly after quoted material containing misspellings, malapropisms, or grammatical errors to indicate that the quote is thus in its original form. *The manuscript guidelines urge us "to carefully examine" [sic] the manuscript for unnecessary split infinitives.* Use brackets with *sic*, as it is an interpolation.

Since. *Since* has a strong temporal meaning (*We have been working on this compound since early in 1987.*) Avoid using "since" when you mean "because." Not *Since there have been problems in synthesizing the compound, we have delayed reporting our results* but *Because of problems in synthesizing the compound, we have delayed our report.*

Single quotation marks (apostrophes). Use single quotation marks for quotations within quotations. If no initial quotation marks are used (for instance, in block style), use double quotation marks for quotes within quotes.

SI units (Système International d'Unités). The SI is a rationalized system designed to correct some confusions and inconsistencies in the traditional metric system. It is based on seven fundamental units: the meter, kilogram, second, mole, ampere, degree kelvin, and candela. (See Table 1.) Other units derive from these base units. Prefixes join base units to express multiples (see Table 2). Factors, powers of 10, are expressed with exponents that are multiples of 3. Avoid *hecto-, deca-, deci-, and centi-*, as they do not represent powers that are multiples of 3.

Journals that have adopted SI permit some widely used units outside the SI system, like *liter, hour, bar,* and *angstrom*.

Use exponents rather than abbreviations with SI reporting: *2 m^2* not *2 sq m*.

Use spaces rather than commas in numbers of five or more digits. If the number has a decimal, use no space: *6000* not *6,000; 25 000* not *25,000;* but *25175.175* not *25 175.175.*

Abbreviate units when they are used with

Table 1. SI Base Units

Property	Base Unit	SI Symbol
Length	Meter	m
Mass	Kilogram	kg
Time	Second	s
Amount	Mole	mol
Thermodynamic temperature	Kelvin	K
Electric current	Ampere	A
Luminous intensity	Candela	cd

Table 2. SI Prefixes

Factor	Prefix	Symbol
10^{18}	exa-	E
10^{15}	peta-	P
10^{12}	tera-	T
10^{9}	giga-	G
10^{6}	mega-	M
10^{3}	kilo-	k
10^{-3}	milli-	m
10^{-6}	micro-	μ
10^{-9}	nano-	n
10^{-12}	pico-	p
10^{-15}	femto-	f
10^{-18}	atto-	a

numbers (*6 g* not *6 grams*). The abbreviations for hour, day, and year are h, d, and y. Minutes, week, and month remain min, wk, and mo. Do not abbreviate units of time except in figures. Do not add an "s" for the plurals of units of measure, unless spelled out (*2 mL* not *2 mLs* but *Two milliliters was added*). Do not use points with abbreviations (*2 mL* not *2 mL.*). Use hyphens for unit or compound modifiers: *a 10-mL flask*. Write out units of measure that begin sentences or titles: not *25 mg was added* but *Twenty-five milligrams was added*. Do not capitalize SI units when spelled out, except in titles. (See also **Abbreviations and acronyms.**)

Regard units of measure as collective singulars: *10 mL is added* not *10 mL are added*.

Leave one space between number and unit (*2 mL*) but not between prefix and unit (*kilojoule* not *kilo joule*).

Slash or solidus. See **Virgule.**

SOPs. See **Instructions, Standard Operating Procedures (SOPs), and procedures.**

Sources. See **References, in text; References, lists of; Tables.**

Species name. Do not capitalize a species name in text or in title. *Frenkel JK, Dubey JP, Miller NL. Toxoplasma gondii in cats: fecal stages identified as coccidian oocysts. Science 1970; 167:893–896.*

Statistics, symbols used in. Common statistical symbols include ANCOVA (analysis of covariance), ANOVA (analysis of variance), $\bar{x}$ (arithmetic mean), D (difference), *df* (degrees of freedom), f (frequency), log (logarithm to base 10), ln (natural logarithm), MANOVA (multivariate analysis of variance), n (sample size), N (population size), P (probability), r (correlation coefficient), R (regression coefficient), RSD (relative standard deviation), R^2 (multiple correlation coefficient) s^2, δ^2 (sample variance), SD,δ (standard deviation of a sample), SE or SEM (standard error of the mean). See also **Mathematical expressions.**

Stet. Proofreading term meaning "let revised text stand as originally presented."

Stubs. See **Tables.**

Subheadings. See **Headings and subheadings.**

Subject-verb agreement. See **Number of subject and verb.**

Subtitles. See **Titles.**

Suffixes. Hyphenate before suffixes that would result in three consonants or two vowels: *cell-like*. Use no hyphen when adding *hood, like, wise, less,* or *fold,* unless the root word is a proper noun. *The twofold increase in recall was accompanied by a threefold increase in error; Feynman-like manipulation.*

T

Table of Contents. Many readers use the table of contents to a manual, report, or book for their first overview of the work. Titles and first-level headings should be as informative as possible. If the readership will be diversified, postpone the use of highly specialized technical terms to second-level headings when practical. For longer works, consider using a general table of contents at the front, and adding detailed, second-level contents before each section or module.

Use type that reflects the division of major and minor headings in the contents; for instance, use larger, darker type for major headings and lighter type for minor ones, or all capitals for major points and initial capitals and indentation for minor points. Keeping the language parallel can also help make a table of contents more readable: mixing phrases and sentences creates a jarring effect.

not

but

Tables. Tables display data concisely and vividly. They are clear, effective summaries; they allow a reader to see relationships and make comparisons; and they inform without interrupting the narrative. While graphs are excellent for trends, tables are ideal vehicles for showing detail.

Tables should be understandable independent of the text, and should supplement rather than repeat it.

Titles should epitomize main ideas or major benefits. They are placed above the table. Set them off with larger type, boldface, or reversal (printing white on black or another dark background rather than black on white). If you use only one table, it need not be numbered. For more than one table, number the series consecutively, and give the table reference on first mention in the text. Do not use a table without a corresponding reference in the text. Most publications use consecutive arabic numerals for tables; some use roman numerals.

Set off column heads in boldface or with shading so that readers can easily see the basic groupings of the data. Abbreviations are permitted in column heads, usually set off by commas. The first word is usually capitalized and the head aligned at the bottom. Experiment with wording since the number of characters affects column width. Use a straddle rule (a rule that goes only the width of the column) to clarify column heads. See also **Column headings.**

Take advantage of the stub column (the column farthest left in the table) to label horizontal rows. Indent within the stub column to show parts of the whole. If column width prohibits indentation of categories, try using smaller type or a variation in type (boldface, lightface) to show the relationship of parts to the whole. If the stub column has a heading, it is called the "stub head." However, if a heading is unnecessary, the space may be left blank.

Use white space or rules to separate horizontal groups in a table. Rules are customary below title and column heads and at the bottom of the table. Avoid unnecessary horizontal and vertical rules, although especially complex tables may need them for clarity. Allow sufficient white space for the reader's eye to find important data. To crowd a table is to impair its purpose.

Align columns by common elements, such as a decimal point or plus/minus sign. Apportion space evenly between columns.

Use unexpanded abbreviations in column heads to conserve space. Use footnotes for information that will not fit in the title or column heads. Use lowercase letters for footnotes, unless the rules of a publication specify otherwise. Avoid numbers, as superscripts are easily confused with numbers in the tables. Place footnotes at the bottom of the table, not the bottom of the page. See **Footnotes.**

If all the information in the column is identical, or if there are no data for most of the entries, the column may be removed and the information placed elsewhere. Do not use columns for data that may be calculated easily from given columns.

Set up primary comparisons horizontally.

Do not use zeros except for a zero reading; use ellipses plus a footnote explaining that no information was available. Prefer decimals to fractions, unless the decimal implies a greater accuracy than that actually achieved.

Do not use ditto marks.

Prefer a space instead of a comma with numbers of five or more digits (*60 000* not *60,000*). See **SI units.**

If you cannot be sure where tables will be placed in a publication, use "See Table 1" or "Table 1" rather than "the table above" or "the table below."

Try to design tables so they can be viewed upright ("portrait") rather than horizontally ("landscape").

Tense of verbs. In a scientific or technical report, the governing tense of the introduction is the present (*We report here data on 18 unselected patients . . .*), although the past tense will also occur in descriptions of procedures (*We examined the potential influence of physiologic amounts of calcium in patients with primary hyperparathyroidism, and we report here data on 18 unselected patients studied at both the lower and upper limits of a normal dietary intake of calcium*).

The governing tense of the procedural section of a paper or report is the past. Do not switch in recounting a procedure from past to present (not *In the protocol, the patient consumed a 400-mg diet on the first day; on the second day, the patient consumes a 1000-mg diet* but *In the protocol, the patient consumed a 400-mg diet on the first day; on the second day, the patient consumed a 1000-mg diet*).

The governing tense of the results section of a technical paper or report is the present (*The data are given . . . results are shown*).

The mean data . . . are given in Table 1, and indi-

vidual values for fasting serum immunoreactive parathyroid hormone, nephrogenous cyclic AMP, and plasma 1,25-$(OH)_2$D are shown in Figure 1. During the restricted-calcium diet, elevated values for serum immunoreactive parathyroid hormone, nephrogenous cyclic AMP, and 1,25-$(OH)_2$D were observed in 17, 17, and 12 patients, respectively.

The governing tense of the discussion is the present tense.

This method provides a means for studying a new class of symmetries (Σ^-, Σ_g^-, A_{2g}, T_{1g}) of molecular excited electronic states; many of these states are metastable or dark states that are likely to act as energy reservoirs in reactive systems such as those found in atmospheric and interstellar chemistry. In addition, the production using this method of such metastable excited states as the $I^1\Sigma^-$ state of CO has possible applications in the study of gas-phase collisional processes and gas-surface interactions, and in the study of higher-lying excited states. We are also investigating analogous identity-forbidden transitions in three-photon absorption.

That is. The phrase is considered parenthetical, and is therefore set off by commas.

That vs. which. Use *that* to introduce restrictive clauses; use *which* for nonrestrictive clauses. *The procedures that are reasonable will be maintained.* (Only the reasonable procedures will be maintained. *That are reasonable* is restrictive or limiting information.) *The procedures, which are reasonable, will be maintained.* (All the procedures will be maintained. *Which are reasonable* is parenthetical, not restrictive.) See also **Nonrestrictive vs. restrictive words, phrases, and clauses** and **Comma.**

Titles. Keep the title specific and informative. Shape the title so that it functions as a preview of the contents that follow. An effective title encapsulates not only the objectives, but the findings and implications.

When possible, avoid formulas in favor of names. Some indexing services cannot handle formulas.

Avoid unnecessary expressions like "A Study of" or "A Determination of" in the title.

Subtitles, usually introduced with a colon, are warranted when they amplify a title (*Type A Viral Hepatitis: New Developments in an Old Disease*). Do not use a subtitle if the information can be included in the title. Not *Six Year Follow-up: The Norwegian Multicenter Study* on *Timolol after Acute Myocardial Infarction* but *Six-Year Follow-up of the Norwegian Multicenter Study on Timolol after Acute Myocardial Infarction.*

Trademarks. See **Brand names and trademarks.**

Transitions. Transitions are words, phrases, or sentences that provide continuity between main ideas and the development of these ideas, between sentences in a paragraphs, or between paragraphs in a longer document. Transitional words showing addition include *additionally, also, and also, again, as well, as well as, besides, equally important, then, first, second, last, further, furthermore, moreover, next.* Transitions showing comparison include *similarly, likewise, in the same way.* Transitions showing contrast include *after all, yet, still, even so, however, in contrast, in spite of that, nevertheless, on the contrary, otherwise.* Transitions of emphasis include *certainly, especially, in fact, to be sure, surely, in particular, indeed.* Transitions of exemplification include *as an example, for example, for instance, as an illustration, in other words, in particular, that is.* Transitions of consequence include *accordingly, as a consequence, consequently, for this/that reason, hence, therefore, thus.*

Trim. Size of the page edge to edge.

Typeface. Several hundred typefaces (designs) exist, among them Gothic, Courier, Helvetica, Times Roman, and Prestige Elite.

08 pt Times Roman	08 pt Helvetica
10 pt Times Roman	10 pt Helvetica
12 pt Times Roman	12 pt Helvetica

Type size. See **Leading, Pica, Point and pitch, Points.**

U

Unit modifiers. See **Hyphen.**

Units of measure. See **Abbreviations and acronyms, Hyphen, Number of subject and verb, SI units.**

Unnecessary words. According to Newton, "more is in vain when less will serve." For a more concise style, trim redundancies (unnecessary repetitions of words or their meanings), inflated phrases, and questionable phrases. Here are some typical examples.

REDUNDANT EXPRESSIONS

interval *of time*	shorter *in length*
two months' *time*	heavier *in weight*
two months' *time duration*	*legal* right
a time span of two months' *duration*	remand, refer *back*
2 a.m. *in the morning,* 2 p.m.*in the afternoon*	*soothing* tranquilizer
twelve noon *twelve* midnight	*free* gift
round, rectangular, oval *in shape*	*future* plan
red, brown, white *in color*	rarely, seldom *ever*
my *personal* view, opinion, preference	repeat *again*
rough, smooth *texture*	*new* addition
rough, smooth, tender *to the touch*	*horrible* tragedy
sour, sweet *tasting*	*utter* chaos
shiny, bright *in appearance*	*different* kinds
skin rash	so *that*
two equal halves	adequate *enough*
all *of*	*proceed to* walk
1 *out* of 10	*set of* twins
each individual	*local* neighborhood
estimated at *about, roughly, more or less*	*contributing* factor
fuse, join, combine, unite *together*	the *present* incumbent
consensus *of opinion*	today's *modern*
in order to, that	today's *contemporary*
fewer *in number*	*contributing* factor

INFLATED EXPRESSIONS

at this point in time, that point in time	now, then

among all of the problems that exist today	among all current problems
due to the fact that	because
owing to the fact that	because
because of the fact that	because
for the simple reason that	because
despite the fact that, in spite of the fact that	although
notwithstanding the fact that	although
regardless of the fact that	although
majoring in the field of economics	majoring in economics
in view of the fact that	as
it is important that	one must, you must
in a very real sense	truly
such as x, y, and z, etc.	such as x, y, and z
under circumstances in which	if
in the event that	if
if it should transpire that	if
if it should transpire in circumstances that	if
as to whether or not	if, whether
has an effect upon, impact upon	affects
make a measurement	measure
has a corrosive effect upon	corrodes
conducts an investigation of	investigates
performs an investigation of	investigates
the question as to whether	whether
whether or not we should measure	if we should measure
used for conservation purposes	used to conserve

QUESTIONABLE EXPRESSIONS

in other words
it goes without
saying that, one

scarcely need
mention that
it is important to
note that
needless to say
it may be said that
it was found that, it
was concluded that
it was demonstrated
that

U.S. See **Abbreviations and acronyms.**

V

Verbs. See **Active vs. passive voice, Collective nouns, Infinitives, Nominalizations, Number of subject and verb,** and **Tense of verbs.**

Verso. See **Recto.**

Virgule (Solidus, Slash). Use the virgule (a diagonal line) to divide numerator and denominator in simple, subscript, and superscript fractions: $3/4$, $q^{1/2}$. Use the virgule for equations that are run into text rather than stacked and displayed. Thus the stacked equation

$$y = \frac{\sin x}{x}$$

becomes $y = (\sin x)/x$.

Use the virgule to mean "per" with abbreviated units of measures. *The speed of light is* 2.99×10^8 *m/s* but *The excimer laser has a repetition rate of 100 pulses per second.* (Pulse is not a unit of measure.) Do not use the virgule with a unit of measure without a specific numerical quantity. Not *We measured the speed of light in meters/second* (no specific numerical quantity is given) but *We measured the speed of light in meters per second.*

Avoid the virgule in "he/she" constructions. Instead, try either a plural or a rewording of the sentence. See **Gender.**

Viruses. See **Nomenclature.**

W

Well, use of hyphen with. Hyphenate "well" in a premodifying position: *It is a well-prepared site* but *The site is well prepared.* Do not hyphenate "well" if it is modified. *It is a very well prepared site.* See **Hyphen.**

Whether. "Whether" introduces alternatives. Avoid the redundant use of "whether or not" in *He did not know whether or not to measure the rainfall.* (Use *He did not know if he should measure the rainfall.*) See also **Unnecessary words.**

Which vs. that. See **That vs. which.**

Widow. A short final line of a column or paragraph appearing at the top of the page.

EXERCISES

1/////

EDIT FOR ORGANIZATION THAT SHOWS MAIN POINTS AND SIGNIFICANCE

■ 1.1 Write an introductory sentence that sums up this paragraph's main idea. If you like, cast it as a question that the paragraph answers. ■

Although our knowledge of the brain is incomplete, most scientists believe that, unlike computers, the brain generally has no specialized memory centers. Instead, there is evidence that information storage involves the synapses—the points at which information is transferred from cell to cell. Synapses are modified in learning situations. Scientists believe that memory storage occurs while the brain is processing information. Unlike standard computers, the brain can analyze information on many channels simultaneously.

■ 1.2 Edit the following section of a report so that it begins with an explicit statement of the main idea and then develops the supporting detail. If you like, cast the main idea as a question. ■

Certain solid substances, notably some crystals and ceramics, exhibit the piezo-electric effect; that is, they generate electric voltage when subjected to mechanical stress.

Conversely, when voltage is applied to such substances, they undergo a mechanical distortion, twisting or bending, expanding or contracting. This distortion can produce a vibration in a certain narrow frequency range characteristic of the substance.

The piezo-electric element in a watch generally has a contact terminal of silver that is about 8 to 20 millimeters in diameter. It sits on top of a very thin film of ceramic and receives a tiny alternating current, converted from the direct current of the watch's battery.

The noise in many electronic alarm watches and clocks relies on the piezo-electric effect, discovered by Pierre Curie.

In watches, a type of ceramic foam is used. The substance resonates at a frequency of around 4,000 hertz, or cycles per second. This produces a sound sharply audible to the human ear, which is most sensitive in the range of 4,000 to 6,000 hertz.

■ 1.3 Edit this section of a report so that the main idea is stated explicitly at the top. ■

Galileo was the first to devise a plan for measuring the speed of light. In the early 1600s he suggested measuring the time it would take a distant person to see the light of a lantern once its shutter was opened. Such an experiment would have failed, however, because Galileo did not have the means to measure the speed of light.

Ole Roemer, a Danish astronomer, reported the first real measurement in 1676. Roemer was studying the eclipses of Jupiter's moons when he noticed that the intervals between the disappearance of some of the moons behind Jupiter varied with the distance between Jupiter and the Earth. Roemer reasoned that the velocity of light caused the eclipse to seem delayed when Jupiter was farther away. Roemer calculated the speed of light to be 140,000 miles per second. The current value for the speed of light is 186,282 miles per second. Roemer's inaccuracy stemmed from his not knowing the precise distance to Jupiter.

A more precise measurement was performed in 1926 by Albert A. Michelson using a rapidly rotating mirror and a beam of light.

Current measurements, which are accurate to within four billionths of a second, are achieved using a laser beam and an atomic clock.

■ 1.4 Edit this memo. Be sure to provide an opening paragraph that summarizes the main point. ■

To: xxxxx   Date:xxx
From: xxxxx   Subject:xxx

The study of sodium bicarbonate appeared last month in *The American Journal of Medicine*. Researchers injected bicarbonate or an innocuous saline solution into patients who had recently suffered heart attacks. The control patients receiving the saline solution had no adverse effects. But among those receiving bicarbonate, researchers found that blood flow decreased, the body's use of oxygen fell about 25 percent, oxygen used by the heart dropped 20 percent, and blood acid levels rose.

Sodium bicarbonate, an alkaline substance more commonly known as baking soda, is widely given to heart attack victims to prevent lactic acidosis, a build-up of damaging acids in the blood. However, researchers have found that solutions of the sodium bicarbonate actually produce difficulties.

Solutions of sodium bicarbonate actually worsen heart and liver functions in patients. Sodium bicarbonate has come under increasing suspicion in recent years, and is not used as routinely as it was in the past. Still, the study is the first on humans to show that it may have harmful effects.

Sodium bicarbonate has been used since the

1920s. The liver normally regulates the body's alkaline-acid balance, but it does not function during a heart attack. It seemed reasonable that if a patient had acidosis, giving a base such as sodium bicarbonate would mark an improvement.

■ 1.5 Edit this memo. Be sure to provide an opening paragraph that summarizes the point. ■

To: L. K. Jones, Quality Control
           20 March 1991
From: R. P. Smith, Microbiological Testing
           Staff
Subject: Microbiological clearance of samples
           for consumer test

Four varieties of sausage from the November 1, 1990, production run at MeatCo's Mason City plant were sliced and vacuum-packed on March 2, 1991, for use in a consumer test. A preliminary microbiological screening was performed on 15 two-ounce packages received by Microbiology Services on March 3, 1991. Tests for the detection of *Staphylococcus aureus, Bacillus cereus*, the Enterobacteriaceae and fecal Streptococci, yeast, and mold were performed by Johnson Laboratories. Composite *Salmonella* and *Listeria* tests were conducted in-house. All samples were deemed microbiologically acceptable. An additional 120 samples (3 per case × 10 cases per variety × 4 varieties) were collected on March 16, 1991, and tested. The standard plate count, fecal Strep., coliform, and *S. aureus* tests were performed by Johnson Laboratories; the composite *Salmonella* and individual *Listeria* tests were performed in-house. All samples were deemed microbiologically acceptable.

The sliced sausage lots 06091, 06092, 06093 and 06094 are cleared for use in consumer test #NPCC 310. However, please note that if the test is not fielded within 4–6 weeks, we strongly suggest in view of the possibility of packaging material delamination and consequent oxygen permeability, that additional samples be submitted to Microbiology Services for mold analysis prior to the product's release. Further, we recommend that all product continue to be stored below 50°F until the test is fielded in order to minimize the potential for growth of microorganisms.

■ 1.6 Below is a trip report in which the author was allowed approximately 140 words for the summary. The abstract is poor; edit it so that it presents an informative summary of the report that follows. ■

TRIP REPORT

Date: Distribution:
From:
Subject: Second International Conference on
         Vacuum Electronics Bath, England,
         July 10–12, 1989

Summary: The conference was an interesting one. The keynote speaker addressed the importance of microvalves, and their probably market position by the mid 1990s. The 18 sessions were well attended and useful.

Background

In the two decades since the solid-state transistor was invented, old-fashioned valves have gradually vanished from the market, except for use in television sets and some military and home electronics applications. Recently, however, the manufacturing technology that let transistors shrink to microscopic size is turning toward valves, or, more specifically, to microvalves.

The newly developed microvalves have many promising features. In the older, fist-sized valves of the 40s, the high speed at which electrons flow through a vacuum offered little advantage because of the large distances they had to travel. The newer valves are the size of a micron. When the microvalve's cathode is subjected to an electric field, which can be varied to control the valve's ability to amplify signals, electrons produced by quantum-mechanical tunnelling arrive at the anode almost instantaneously. The speed at which the electrons pass through each microvalve's vacuum makes

microvalves the fastest electronic technology. They have another advantage besides speed. Because of their size, vacuum pressures might not need to be as low as in older valves. This could help make it quite easy to manufacture microvalves.

Because of its vacuum, the valve is far less susceptible to radiation and voltage surges than transistors. Surges in voltage, high temperatures and other high pulses of energy can disable chips by making electrons jump between parts of their circuits. Too high a voltage applied to a solid-state chip creates a permanent conducting channel between its terminals, destroying it. At high temperatures, solid-state chips conduct electricity too readily, and chips in satellites are disturbed by bands of natural radiation around the earth. The pulse of electromagnetic radiation caused by a nuclear explosion would destroy weapons and communications systems that use chips.

Dyniomac's Work

Dyniomac (Britain) runs the world's biggest vacuum microelectronics research program. It usually makes microvalves by taking a sliver of silicon and chemically etching millions of two-micron-high conical metal cathodes on it. A microscopic film of silicon dioxide is then put over the cone-covered silicon, followed by molybdenum or another metal. This metal layer, when subjected to an electric current, produces the electric field that activates the microvalve's cathode. One-micron holes are then etched through the top two layers around the cathodes to make a grid of tiny vacuum chambers.

It is technically challenging to produce huge numbers of the tiny cathodes consistently enough to emit electrons in a stable and identical fashion. Researchers reported that the stability of emissions seems to improve when the cathodes are coated with a microscopic film of fatty acid. They have also found that microvalves perform less erratically if they are cleaned by heating during manufacture.

Military Research Laboratory

Researchers at the Military Research Laboratory (US) think the microvalves may have applicability for high definition television (HDTV) screens. Instead of one electron beam scanning across a screen, the television would use a microvalve display of millions of tiny cathodes, each firing a single beam onto a phosphor-coated glass plate. This would produce higher contrast, better color, and no deterioration in picture quality when viewed from the side. The picture would appear instantaneously as microvalves need no time to warm up. Researchers there imagine huge, flat-panel HDTVs in every living room by the late 1990s. By this time, microvalves may be considerably cheaper to produce than rival flat-screen components.

Conclusions

The 18 sessions, spread over three days, addressed the latest developments in a new generation of microvalves. The research findings of two major presenters are summarized above. The well-attended conference attracted representatives from 12 countries. Nearly half the attenders were from companies (142), half from academic centers (150).

Operating at up to 20 times the speed of the best solid-state chips, the new valves are fast and highly compressed: A million fit on a square centimeter of silicon. By 2000 these microvalves could be as cheap to make as today's chips. While no one expects vacuum microelectronics to obviate solid-state chips, they may replace them in some applications where speed and ruggedness matter. Commercial versions are unlikely to surface until the mid-1990s, most probably replacing chips as sensors in hot places, like the inside of jet engines or inside nuclear reactors. They could also displace cathode-ray tubes in some applications where smallness and durability count.

■ 1.7 The following selection is from a handbook on animal health written for livestock and pet owners. Edit the abstract that introduces the selection so that it is informative. Assume a 140-word limit. ■

Abstract: This section discusses two common diseases among dogs and cats: rabies and toxoplasmosis. Background and guidelines for treatment are offered.

Rabies

Rabies, one of the oldest recorded diseases in history, is caused by an RNA virus. The virus can be found in nerve tissue, in saliva and salivary glands, in the pancreas, and in the urine and other body fluids of infected animals. Rabid dogs and cats are the main source of human infections. Wild animals such as rabid skunks, foxes, raccoons, coyotes, bats, and bobcats are other sources of infection.

Most commonly rabies is spread by the bite of an infected animal, through the presence of the virus in the saliva. Broken skin of the bite wound allows the rabies virus to get into the body, where it can flourish.

Once inside the body, the virus is drawn into the nerves and follows nerve fibers to the brain and salivary glands. Brain lesions lead to altered behavior, aggressiveness, progressive paralysis and, in most species, death. Once the virus arrives at the salivary glands, the infected animal can spread the virus by way of its contaminated saliva to susceptible animals.

Signs of the Disease

Not all rabid animals show the same signs of the disease. Some animals will ultimately show the classic behavior of the mad dog, while others may withdraw quietly to a dark, sheltered place. Other than altered behavior states, there are no obvious signs that an animal is infected.

Any unprovoked attack by an animal should suggest the possibility of rabies. Dogs and cats that are rabies suspects should be confined to a cage where they can be observed and fed without risk.

Diagnosis

No acceptable diagnostic tests can be used to evaluate a live dog or cat. All laboratory tests for rabies are presently done on the dead animal.

The preferred technique is to use fluorescent antibody tests and to inject brain tissue of the suspect animal into the brains of mice. Rabies virus will cause the death of the injected mice. The inoculation test of mice is quite sensitive.

Treatment

Although exposure to rabies is a cause for concern, when prompt action is taken there is no cause for panic. If you or someone with you is bitten by a rabid animal, you should wash the wound vigorously with soap or detergent and flush it repeatedly with large amounts of water. Contact a physician as soon as possible about the wound and the type of animal that attacked. Contact a veterinarian or local animal control officer and report the attack. Keep track of the animal involved in the attack. Responsible authorities will either quarantine the animal for observation or kill the animal and check for the rabies virus.

To reduce exposure to rabies, vaccinate all your dogs and cats. Rabies vaccines produced today for use in dogs and cats are both safe and effective. Regular rabies vaccination every one to three years should be a part of any responsible pet health program.

No vaccines have been tested or approved for use in any wildlife species. Vaccination of wildlife species may cause rabies or death of the animal.

Toxoplasmosis

This disease is caused by infection with *Toxoplasma gondii* and occurs throughout the world. Infection has been observed in a wide range of birds and mammals.

Toxoplasma organisms living in body cells of the host cause illness when they escape these body cells. Cats are the primary host, but other mammals and birds may also be infected.

Animals acquire the toxoplasma organism by eating infected raw meat or by ingesting contaminated feces. Some species can also acquire the toxoplasma infection during pregnancy, infecting the developing fetus.

Human infections with *Toxoplasma gondii* can occur both prenatally and postnatally. The two postnatal modes of transmission are by the ingestion of infected raw or undercooked meat, or feces.

Inadvertent ingestion of organisms in feces is usually related to contact with cat litter boxes or contaminated soil. Prenatal transmission occurs when a woman acquires the infection during pregnancy. Fetuses appear to be at great risk to toxoplasmosis during late pregnancy.

The great majority of toxoplasmosis infections causes no apparent illness. The most frequent signs of illness in cats and dogs are associated with infections to the nervous system, eyes, respiratory tract, and gastrointestinal system.

Cats may experience fever, jaundice, enlarged lymph nodes, difficulty in breathing, anemia, eye inflammation, abortion, encephalitis, and intestinal disease. Cats may also develop stiff, painful muscles from the infections so that they are unable to move.

Pneumonia, liver disease, and ocular, nervous system, and muscle damage may result as signs of illness in the dog.

Diagnosis of toxoplasmosis based on history and signs of illness alone is usually not possible because of the wide variety of signs of illness that can occur. Identification of toxoplasma organisms in the feces of infected cats is possible in early infections but the organisms are excreted for one to two weeks only.

Laboratory tests currently used in the detection of *Toxoplasma gondii* antibodies are generally preferred to the identification of organisms in feces. Presence of toxoplasma antibodies suggests that the animal may be immune to infection.

Lack of antibodies usually indicates the animal is susceptible and could shed or be shedding organisms in the feces.

Avoiding Infection

Some recommendations can be made to prevent ingestion of infected meat and to minimize exposure to contaminated feces.

To prevent infections of cats and other animals, confine the animals to their home environment, avoiding exposure sources. Feed animals only commercial or well cooked meat, and never feed raw meat to cats or dogs.

Change cat litter boxes frequently and dispose of the feces so that no animals or person will come in contact with the feces.

For human consumption, cook meat thoroughly to destroy any organisms present. Always wash hands thoroughly after handling meat.

Wear gloves when gardening, especially in areas favored by cats for defecation, and when changing cat litter boxes. Cover children's sandboxes when not in use so cats cannot defecate in them. Encourage children to wash hands thoroughly before eating.[15]

2/////

EDIT FOR TITLES,
HEADINGS, AND SUBHEADINGS
THAT MARK MAIN POINTS
AND THEIR ORDER

■ 2.1 The following selection is from an informative booklet on causes and prevention of food poisoning. This excerpt discusses causes of food poisoning. (1) Write an opening or lead paragraph that prepares the reader for the information that follows. (2) Add a section heading and subheadings to show the reader the organization of the excerpt. ■

Staphylococcus aureus is the small, round organism that is a leading cause of food poisoning. We literally carry staph with us, for it lives in our noses and on our skin. It is found in concentrated form in boils, pimples and other skin infections.

When transmitted to food, usually by handling, staph starts growing. At warm temperatures, certain types of staph multiply rapidly and produce a toxin or poison that makes people sick. Nausea, vomiting and diarrhea usually appear 2 to 6 hours after eating staph-infected food, and last a day or two. The illness is usually not too serious in healthy people.

Salmonella, which appears as short, thin rods under the microscope, is another major cause of food poisoning. Actually, salmonella is the name used for some 2,000 closely related bacteria that cause more severe flu-like symptoms than staph: diarrhea, vomiting, and fever.

Salmonella continually cycles through the environment in the intestinal tracts of people and animals. Salmonella is often found in raw or undercooked food, such as poultry, eggs, and meat.

Perfringens, full name *Clostridium perfringens*, ranks third as a cause of food poisoning. It, too, is present throughout the environment—in the soil, the intestines of animals and humans, and in sewage. Perfringens differs from staph and salmonella, however, in two ways. First, it's anaerobic, which means it grown only where there is little or no oxygen. Second, it produces two kinds of cells.

The normal perfringens cell is the unpleasant one: it produces the poison that makes you sick. But perfringens has a spore cell, too, that can survive circumstances that eliminate normal cells. These spores are tricky, because at temperatures between 70° and 120°F, they can become normal cells again, multiplying quickly to disease-causing levels.

Perfringens shows an ugly side—usually diarrhea and gas pains—some 8 to 24 hours after consumption. While the symptoms often end within a day, people with certain medical conditions such as ulcers can be seriously affected.

Botulism, while very rare, is the deadly food poisoning caused by *Clostridium botulinum*. Although it needs just the right conditions to develop, botulism is clearly a danger because the spores are always around in soil and water.

Like perfringens, the botulinum bacteria grow best in anaerobic (reduced oxygen) conditions. Since the canning process forces air out of food, the botulinum bacteria may find improperly canned foods a good place to grow.

Low-acid vegetables such as green beans, corn, beets, and peas, which may have picked up botulinum spores from the soil, are at risk. The risk is greater if they are home-canned, and safe canning procedures have not been followed precisely.

Unfortunately, the bacteria that cause food poisoning are not, like food spoilage organisms, obvious. The bacteria that cause food poisoning, with its mild-to-severe intestinal flue-like symptoms, can't be seen, smelled, or tasted.[16]

■ 2.2 The following selection is from the same booklet on causes and prevention of food poisoning. This excerpt discusses special care for special foods. (1) Write an opening or lead paragraph that prepares the reader for the information that follows. (2) Add a section heading and subheadings that show the organizational pattern. ■

Hamburger receives more handling than many other meats. The beef is butchered and then ground. Trimmings from more expensive cuts and small amounts of fat may be added to the mixture for moisture and flavor. Hamburger is thus exposed to many of the common food-poisoning bacteria, including salmonella and staph.

Hamburger can give you trouble if you eat it raw or rare. For complete safety, make sure hamburger is *brown or at least brownish pink in the center* before you serve it.

If you are making a meatloaf, use a meat thermometer to make sure meat cooks to 170°F. This is particularly important if your mixture contains pork.

Many people have questions about buying and storing hamburger. Generally, hamburger at the store should be bright red to dullish brown. Return any package that has an off-odor when opened.

You can store hamburger in the coldest part of the refrigerator for use in a day or two.

Otherwise freeze it. It will keep frozen at full quality for 3 to 4 months.

Another special food is ham. Because ham is cured—and often smoked, aged, and dried—you may think it is "protected" against food poisoning. It isn't always. Ham, like all pork, can spread food poisoning bacteria—chiefly staph and salmonella.

The thing to remember with ham is to read the label carefully. Know exactly what kind of ham you've bought, and then observe these guidelines:

For refrigeration, ham slices or whole hams bought in paper or plastic wrap should be stored in the coldest part of the refrigerator. Ham slices should be used in 3–4 days; a whole ham within a week. Even most canned hams should be refrigerated. Read the label for storage time.

For freezing, remember that freezing ham is tricky. Like other smoked pork products, ham tends to lose flavor and texture in the freezer. To freeze, wrap ham tightly in freezer paper or use special plastic freezer bags. Don't try to keep it frozen for more than a month or two.

For cooking, it is important to read the label before serving ham. "Fully cooked" hams have been completely cooked during processing. They can be served as is, warm or cold. Fresh hams or those labeled "Cook before eating" must be cooked to a uniform internal temperature of 170°F.

We also need to consider mayonnaise. Don't even try to freeze mayonnaise. After opening, place mayonnaise, which keeps best at 50°F in the warmest part of the refrigerator, on a shelf farthest from the freezing compartment. For best flavor, use within two months.

A final word—mayonnaise is not a villain. Contrary to what you may have thought, adding mayonnaise to food does not increase the risk of food poisoning. In fact, most commercially prepared mayonnaises and salad dressings contain lemon juice or some other acid flavoring, which slows bacterial growth. Salt in mayonnaise also retards bacterial growth.

Next is the example of turkey, chicken, and duck with stuffing. Fixing poultry with stuffing gives food poisoning several opportunities to strike. Bacteria present in raw poultry can get into the stuffing. The stuffing, deep inside the bird, may not heat thoroughly to bacteria-killing temperatures. Here are some safety tips:

If you mix your stuffing a day ahead, pre-mix only the dry ingredients and refrigerate them separately from the uncooked bird. This will keep any bacteria in the raw poultry from entering the starchy dressing.

Stuff the bird just before you're ready to cook it, and stuff loosely. This gives heat from the oven a better chance to cook the stuffing all the way through.

Place the stuffing in a separate bowl for serving. Keep the poultry meat and stuffing separate for refrigeration, too.[17]

■ 2.3 The information in the following excerpt is from a booklet on food poisoning. This section describes the three ways to keep a safe kitchen. Improve this excerpt by (1) editing the lead paragraphs to sharpen their focus and (2) supplying subheadings to show the divisions of the section. ■

A Safe Kitchen

Staph, salmonella, perfringens, and the botulinum bacteria are the four main food poisoners. In addition, there are twenty other organisms that can cause problems.

To get food on the table safely, you need to know and follow the rules for food care. The first topic is keeping food hot.

High food temperatures (165° to 212°F) reached in boiling, baking, frying, and roasting kill most food poisoning bacteria. If you want to delay serving cooked food, though, you have to keep it at a holding temperature—roughly 140° to 165°F. Steam tables and chafing dishes are designed to maintain holding temperatures. But they don't always keep food hot enough. So it's not wise to leave hot food out for more than 2 hours.

When cooked food is left out unheated, the possibility of bacterial growth is greater, since the food quickly drops to room temperature where food poisoners thrive.

To serve hot foods safely—particularly meat and poultry, which are highly susceptible to food poisoning—follow these rules:

Cook thoroughly. Cook meats and poultry to the "doneness" temperatures given in your cookbook. To make sure that meat and poultry are cooked all the way through, use a meat thermometer. Insert the tip into the thickest part

of the meat, avoiding fat or bone. For poultry, insert the tip into the thick part of the thigh next to the body.

Don't interrupt cooking. Cook meat and poultry completely at one time. Partial cooking may encourage bacterial growth before cooking is complete.

Thoroughly reheat leftovers. Cover leftovers to reheat. This retains moisture and guarantees that food will heat all the way through.

Never leave food out over two hours.

The next topic is keeping food cold.

The colder food is kept, the less chance bacteria have to grow. To make sure your refrigerator and freezer are giving you good protection against bacterial growth, check them with an appliance thermometer. The refrigerator should register 40°F or lower. The freezer should read 0°F or lower.

Here are some tips for keeping meat, poultry, eggs, milk, cheese, and other perishable foods cold:

If you are shopping, pick up perishables as your last stop in the grocery and get them home and into the refrigerator quickly, especially in hot weather.

Since repeated handling can introduce bacteria to meat and poultry, leave products in the store wrap unless the wrap is damaged when refrigerating.

While "freezer burn"—white, dried-out patches on the surface of meat—won't make you sick, it does make meat tough and tasteless. To avoid it, wrap freezer items in heavy freezer paper, plastic wrap, or aluminum foil. Place new items to the rear of the freezer and old items to the front so that they'll be used first. Dating freezer packages also tells you what to use first.

When you are thawing food, the safest way to thaw meat and poultry is to take them out of the freezer and leave them overnight in the refrigerator. Normally, they will be ready to use the next day.

For faster thawing, put the frozen package in a watertight plastic bag under cold water. Change the water often. The cold water temperature slows bacteria that might grow in the outer, thawed portions of the meat while the inner areas are still thawing.

If you have a microwave oven, you can safely thaw meat and poultry in it. Follow the manufacturer's directions.

It's not a good idea to thaw meat and poultry on the kitchen counter. Bacteria can multiply rapidly at room temperatures.

As to storing leftovers, don't cool leftovers on the kitchen counter. Put them straight into the refrigerator. Divide large meat, macaroni, or potato salads and large bowls of mashed potatoes or dressing into smaller portions. Food in small portions cools more quickly to temperatures where bacteria quite growing.

The third topic is how to keep food safe and clean.

When you shop, be careful in your selection of perishable foods. Make sure frozen foods are solid and that refrigerated foods feel cold. The "Sell by" and "Use by" dates now printed on many products can also be helpful, but you can't rely on them absolutely. They don't reflect a number of things that can shorten a food's useful life, such as too much handling by store employees and customers, or inadequate refrigeration. Therefore it's best not to store fresh meat on the refrigerator shelf unless you plan to use it in a day or two.

The final concern in the home care of food, of course, is keeping food clean. Here are some storage and cleanliness guides.

Store foods in safe places. Store frozen foods in the freezer, perishable food to be used within a few days in the refrigerator, and canned foods in a clean, dry place.

Keep pets, household cleaners, and other chemicals away from food. Don't store food near leaky pipes or seeping moisture. Control household pests (rats, mice, roaches).

Don't spread infection. Always wash your hands before beginning food preparation. Teach this simple but vital rule to your children, too.

Use gloves to handle food if you have any kind of skin cut or infection on your hands. Try not to sneeze or cough into food.

Keep washing and drying cloths clean. Bacteria can loiter in towels and cloths you use over and over, so wash kitchen linen often. Throw out dirty or mildewed dish sponges.

Wash hands, countertops, and utensils in hot, soapy water between each step in food preparation. Bacteria present on raw meat and poultry can get into other food if you're not careful to wash everything they've touched before exposing another food to the same surfaces and utensils. Starchy foods and those con-taining dairy products are particularly vulnerable.

Second, wash your hands, utensils, and food-contact surfaces between contact with raw meat or poultry and the same dish when cooked. For instance, if you use a serving dish to marinate raw chicken, wash the dish well before using it to take up that same chicken after it's cooked.

■ 2.4 The page design of these instructions is poor. Edit the text so that the steps in the sequence are easier to see. ■

Feeding Computer Forms

If computer forms become misaligned during copying, the machine will stop automatically and the message "Misaligned" will flash. To correct this misfeed, follow this procedure. Lift the cover and check the condition of the computer forms. Place the first uncopied page on the glass so that the perforations on the left-hand edge line up with the registration guide. Be sure the forms are lying straight across the document glass. Carefully lower the cover. Press the start button to continue.

■ 2.5 Insert subheadings that will divide this task visually for the reader. ■

Using the Video Machine

1. Turn television "on."
2. Turn channel selector to "Channel 3."
3. Turn video "on."
4. Load cassette into video, matching arrows on the cassette to arrows on machine. It will snap into place.
5. To view tape, press "play."
6. To review material when doing the text/computer exercises, use the rewind, fast forward, or pause button as needed. Note the time-limit of two minutes on the pause function.
7. If you need to pause for more than two minutes, "eject" the tape when doing text/computer exercises.
8. Reload cassette when ready to continue.
9. Press "play" and continue.
10. When you are finished, rewind the tape.
11. Turn VCR off, using "power switch."
12. Turn TV "off."

■ 2.6 In the following instructions, the information is correct but the page layout is poor. Edit for page design.

1. Use headings to group tasks.
2. List steps separately.
3. Make sure all cautions, warnings, and notes are visually distinct. ■

Instructions for Cleaning the Printer

If you are dissatisfied with print quality, clean the inside of the printer by wiping off visible toner with a damp cloth. Four areas of the printer should be cleaned: the transfer wire, the transfer guide, the transfer tray, and the paper guide. Caution: Turn the printer off before you clean it.

To clean the transfer wire, turn the printer off. Open the printer's top cover. Locate the small brush attached inside the printer beside the guide. Using the brush, clean the top, bottom, and sides of the transfer wire. Note: If the brush is not available, use a cotton swab dipped in isopropyl alcohol or water. Make sure the swab does not drip on the rollers or plastic parts.

To clean the transfer guide, turn the printer off. Open the printer's top cover. Wipe the silver strip of the transfer guide with a damp cloth. Caution: Use water only.

To clean the transfer tray, turn the printer off. Open the cover of the transfer tray. Wipe any paper dust off the tray, using a damp cloth.

To clean the paper guide, wipe the paper feed guide with a damp cloth. Caution: Use water only.

■ 2.7 Improve the page design of these instructions by
    1. Using headings to group tasks.
    2. Listing steps separately. ■

Instructions for Using the Phone System

To answer incoming calls, pick up the handset. Press the line key to the right of the flashing red light.

To put a call on hold, press the hold key once. If you don't get back to the call within 45 seconds, a bell reminds you that someone is on hold. Pick up the handset, tell the person to continue to hold, and press the hold key again.

To place an outgoing call, press any line key except ones with red lights either steadily lit or flashing. Key the number. As soon as someone answers, pick up your handset and begin to speak. If you speak before picking up the handset, the person you are calling will not hear you.

To re-call the number you most recently keyed, press any line key except ones with red lights either steadily lit or flashing. Press the # key. The number you most recently keyed is entered automatically.

To place an outgoing call to a pre-programmed phone or beeper number, press any line key except one that has a red light steadily lit or flashing. Press the * key. Key the code for the desired phone or beeper number. If you are beeping someone who does not have a digital display beeper, hang up when you hear several tones following the sound of a normal ringing phone. If you are beeping someone who has a digital display beeper, listen for three tones. Press the * key and then the # key. Key the phone number. Hang up when you hear several tones following your entry of the phone number.

■ 2.8. Edit the following instructions. ■

Instructions for Use of a Mousetrap

Hold the trap in your right hand. Pull the clamp back to the opposite side of the trap. Before you do this, be sure to bait the trap. Use a substance that will attract the mouse, like cheese, peanut butter, or gum drops. You should also be careful to keep your fingers away from the side opposite the clamp once you've pulled the clamp back, as the clamp can return quickly to the platform and hurt your fingers. Once you've pulled the clamp back to the opposite side of the trap, hold it down with your thumb. With the other hand, slide the metal retaining rod over the mousetrap. You should make sure that the rod slides into the trigger before letting go. Place the trap in the path of mice, once you have gently released the tension on the rod. Hold the trap carefully when you move it, as the clamp can return to the platform, powered by the spring, and hurt your fingers.

■ 2.9 Edit this displayed list from an informative booklet. ■

What Are the Exclusive Rights of a Copyright Owner?

According to Section 106 of the 1976 U.S. Copyright Law, as the owner of the copyright you have the exclusive right to do and to authorize any of the following:

1. to reproduce the copyrighted work in copies or phonorecords;

2. to prepare derivative works based upon the copyrighted work;

3. to distribute copies or phonorecords of the copyrighted work to the public by sale or other transfer of ownership, or by rental, lease, or lending;

4. in the case of literary, musical, dramatic, and choreographic works, pantomimes, and motion pictures and other audiovisual works, to perform the work publicly; and

5. in the case of literary, musical, dramatic, and choreographic works, pantomimes, and pictorial, graphic, or sculptural works, including the individual images of a motion picture or other audiovisual work, to display the copyrighted work publicly.

3/////

EDIT FOR DEFINITIONS CRUCIAL TO READER UNDERSTANDING

■ 3.1 Edit for errors in definition. ■

Two techniques can be considered to remove radon from water. The GAC method has been more widely tested and is more commonly used in individual homes. Granular activated carbon removes radon from the water. The second method requires storing water in a tank until the radon has decayed.

■ 3.2 Edit for errors in definition. ■

In some industries earmarked by Japan's MITI as top priorities for development—communications, computers and microelectronics—Japanese companies have come to dominate the global market. Many people attribute Japanese industrial success to the Ministry of International Trade and Industry's policies of establishing long-term goals and pooling resources.

■ 3.3 Edit for errors in definition. ■

What constitutes a user-friendly interface? Menus evolved from early uses in defense computing in the 1960s to the present sophisticated "tear-off menus" the user can move to a convenient spot on the screen and fix there. Windows have also been quite important, from their beginnings in Ivan Sutherland's "Sketchpad," 1962, with its tiled windows, to overlapping and overlaid windows. Windows and menus, two essentials today's PC users take for granted, actually developed slowly as designers sought devices that would make computers easier to use.

Menus are functions continuously listed onscreen that can be called into action with key combinations. Menus list choices. A menu is a list of command options. Some stay on screen. "Pop up" and "pull down" menus are requested by the user. Tiled windows are laid out side by side. For instance, one window can show a view of a ignition system, another window a close-up of a part of the ignition system. Overlapping windows can be stacked on top of each other, or overlaid. A window is an area of a computer display in which a program is executing.

■ 3.4 Edit for errors in definition. ■

The market for flat-panel displays is huge. Designers, manufacturers and users value them for their thinness, durability, and low-power demands. Among LCDs, the TN, STN, and DST dominate the market. STNs are beginning to replace TNs in applications requiring more information. Where greater contrast ratios are required, the DSTs are starting to replace the STNs. Contrast ratio is an important consideration in choosing a flat-panel unit.

Many twisted-nematic LCDs channel light from a fluorescent lamp at the rear of the display through a polarizer. Their nematic crystals are aligned in a spiral configuration that twists the polarized light by 90 degrees. A second polarizer, whose orientation coincides with the light's rotated axis, passes the light, which appears as an illuminated pixel. When an electric field is applied to the nematic crystals, they are untwisted. The angle of the polarized light is not shifted, and light is blocked by the second polarizer, showing upon the screen as a dark pixel.

Nematic crystals are the most commonly used of three classes of liquid crystal. Contrast ratio is the ratio of maximum to minimum luminance. (A contrast ratio of 10:1 or more is needed for best readability.) A liquid crystal display is a display made of material whose reflectance or transmittance changes when an electric field is applied. In the construction of the panel, the orientation of the buffing direction of the front glass plate can be varied from that of the rear glass plate to produce a display with a "twisted" structure. The amount of twist may vary between 90 and 270 degrees; hence the terms "twisted," "supertwist" and "double-supertwisted."

The two leading LCD types are the twisted-nematic LCD and the **double-layer super-twisted-nematic LCD.** The TN LCD is the basic type. Twisted nematic molecules align on a helical axis in the absence of an electric field and usually twist polarized light up to 90 degrees, channeling it on its new axis through an exit polarizer. In the double-layer supertwisted nematic LCD, a variation on the twisted-nematic LCD, liquid crystal molecules are aligned so that they twist beyond 90 degrees as high as 270 degrees. The greater twist allows more lines to be displayed.

■ 3.5 The excerpt that follows is part of the final draft of a booklet on ground water. The author wants the booklet to be read by people planning to buy or build a house. As you read the selection, preparing to edit it for clarity of definitions, put yourself in place of the intended reader. Are all the key terms understandable? If not, when would definitions, explanations, examples, or figures be helpful? Mark the places where terms should be clarified. Instructions follow at the end of the excerpt. ■

Introduction

When buying a home in the country, people need to consider certain factors that usually do not confront the urban home buyer, such as whether the water supply is adequate and if the means of disposing of wastewater is safe. Disappointed rural homeowners have sometimes found out too late that the well drilled on their new land does not yield enough water, or that the water is of poor chemical quality. Foundations can become unstable from excess surface runoff or from high ground-water levels. Septic systems can fail. Wells can be contaminated by septic systems or barnyard wastes. Shallow or dug wells on farms or near older homes that served adequately in earlier years are often inadequate for modern uses.

Preventing water problems or coping with them when buying or building a home can be complex. This booklet describes the most common well problems encountered by rural homeowners, how to recognize them, solve them, or get help. But first, the characteristics and behavior of ground water and the relationship between ground water and the surrounding land are discussed briefly.

The Hydrologic Cycle

The **hydrologic cycle** is the continuous circulation of water from land and sea to the atmosphere and back again. Water evaporates from oceans, lakes, and rivers into the atmosphere. This water later precipitates as rain or snow onto the land where it either

. evaporates
. runs off into streams and rivers
. infiltrates (seeps) into the soil and rock.
Some of this water is transpired back into the atmosphere by plants.

The remainder becomes ground water, which eventually seeps into streams or lakes from which it evaporates or flows to the oceans.

Ground Water

Ground water is that part of precipitation that infiltrates through the soil to the water table. The unsaturated material above the water table contains air and water in the spaces between the rock particles and supports vegetation. In the saturated zone below the water table, ground water fills in the spaces between rock particles and within bedrock fractures.

Where Ground Water Occurs

Rock materials may be classified as consolidated rock, often called bedrock, and may consist of sandstone, limestone, granite, and other rock, and as unconsolidated rock that consists of granular material such as sand, gravel, and clay. Two characteristics of all rocks that affect the presence and movement of ground water are porosity and permeability.

Consolidated rock may contain fractures, small cracks, pore spaces, space between layers, and solution openings, all of which are usually connected and can hold water. Most bedrock contains vertical fractures that may in-

tersect other fractures, enabling water to move from one layer to another.

Unconsolidated material overlies bedrock and may consist of rock debris transported by glaciers or deposited by streams. Well-sorted unconsolidated materials can store large quantities of ground water; the coarse materials—sand and gravel—readily yield water to wells.

Aquifers

Although ground water can move from one aquifer into another, it generally follows the more permeable pathways within the individual aquifers from the point of recharge to the point of discharge. Where water moves beneath a layer of clay or other dense, low-permeability material, it is effectively confined, often under pres-

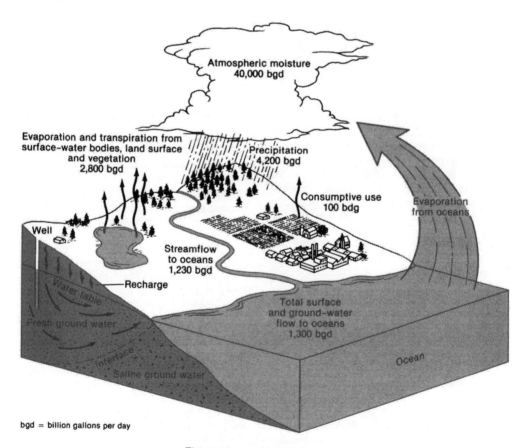

bgd = billion gallons per day

The continuous hydrologic cycle

Fig. E.1. Hydrologic Cycle. *Source:* U.S. Department of the Interior.

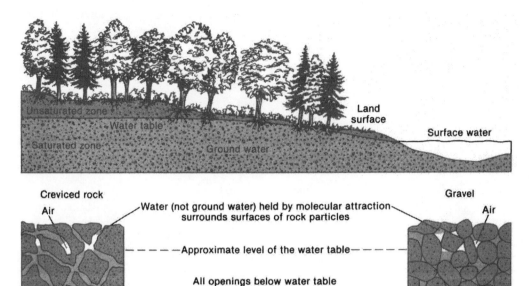

How ground water occurs in rocks

Fig. E.2. How Ground Water Occurs in Rocks. *Source:* U.S. Department of the Interior.

sure. The pressure in most confined aquifers causes the water level in a well tapping the aquifer to rise above the top of the aquifer. Where the pressure is sufficient, the water may flow from a well.[18] ■

Have you marked places in the excerpt where you think definitions would improve the reader's understanding? To help you clarify or expand information in these places, here is a selection of background information to use as a

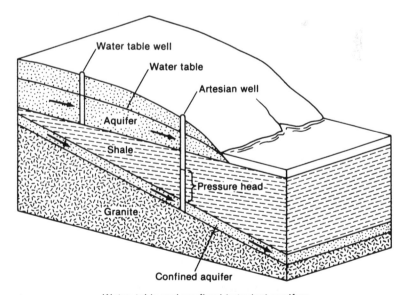

Water-table and confined (artesian) aquifers

Fig. E.3. Aquifers. *Source:* U.S. Department of the Interior.

reference or backup. Return to the excerpt from the booklet and edit it for clarity of definitions, selecting and adapting background information from the following data.

Porosity: the size and number of void spaces.
Permeability: the relative ease with which water can move through spaces in rock.
Aquifers: Most of the void spaces in rocks below the water table are filled with water. Wherever these water-bearing rocks readily transmit water to wells or springs, they are called aquifers.
The point of recharge: areas where materials above the aquifer are permeable enough to permit infiltration of precipitation to the aquifer.
The point of discharge: areas at which the water table intersects the land surface and water leaves an aquifer by way of springs, streams, or lakes and wetlands.

4/////

EDIT FOR A CONCISE, CONSISTENT STYLE

■ 4.1 Edit these sentences for style. ■

1. The fact that three challenges of Supreme Salad Creme, which have the same formulation (22.5% soybean oil, 3.5% distilled vinegar, and 0.4% lactic acid) but were produced at different dates, are behaving differently was discussed.

2. It is imperative, however, in such cases that we obtain documentation that either clearly shows that the pathological process worsened, thus expediting the need for surgical intervention or determine if the planned surgery was performed after entering the study simply because the patient was on a waiting list or; although the indication for the surgery may have been clear to the treating physician before the patient entered the study, the patient did not agree to surgery until after starting the study. The latter two examples demonstrate situations that disqualify a case as reportable and if follow-up reveals either of these situations, the report should be a candidate for deletion.

3. Rest, which reduces the pain, superimposed weakness, and occasional loss of fine motor control found in acute overuse injury, as well as the low-grade symptoms of tenderness and possible swelling, is the cornerstone of therapy.

4. A replacement which may highlight the importance of this position in interacting with the enzyme would be to substitute the ether linkage with a methylene.

5. In a small set of case studies, the basis of findings on running as a possible analogue of anorexia published in the February 1983 issue of the New England Journal of Medicine, the authors, Dr. Yates, Leehey and Shisslak, state that distance running has resulted in a billion-dollar industry for sports goods and numerous sports clinics to treat serious ailments resulting from running, while at the same time anorexic cases have increased to become a public health hazard, a condition they seek to compare and correlate.

6. Common LISP is a dialect of LISP that originated to achieve some degree of commonality so that it could be run on a broad range of platforms as opposed to several other implementations of LISP that began to diverge because of the difference in the implementation environments (e.g., ZETALISP for microcoded PCs, S-1 LISP for supercomputers).

7. Finally, we did eliminate the possibility that the higher number of SPEE-positive samples than SPAA-II positives was not due to the fact that the SPAA-II assay was less sensitive than the SPEE assay. The data show that the SPAA-II assay is at least 50 to 100 times more sensitive than the SPEE assay.

■ 4.2 Edit this memo for style. ■

To: Dr. S. St. James 26 October 1991
From: Dr. R. Smith
Subject: Laboratory Service Assistance

During the time period April 3–7, 1991, we experienced some unusual problems. During that time, we found much of the laboratory maintenance work unsatisfactory for the following reasons:

- On Monday, 8 out of 10 of the glass bins were not emptied. By 4:00 p.m. Tuesday afternoon, the glass trash cans were still not emptied until I asked to have them emptied. Then a liner wasn't put back in so that the trash can could be used. Also, the new biohazard drums were placed in the hall despite the fact that we've asked regularly for them to be stored in the laboratory.
- On Wednesday, different kinds of dirty glassware continued to stack up in piles. Biohazard drums needed to be disposed. The liner for the trash can still wasn't replaced.

Laboratory services knows the routine for our lab. But yet during this past week, I found myself doing so much check up to see if something was done right. Yet glassware was still being placed in the wrong places. Our laboratory service person was not available during the course of the day to ask for things. Because of this fact, the lab was carelessly maintained, instead of being run properly.

■ 4.3 Edit these instructions for style. ■

POLYMER LABORATORY SAFETY

The following instructions must be adhered to at all times.

1. Safety goggles must be worn during all chemical experiments.
2. No smoking and eating are permitted in the laboratory.
3. One should familiarize oneself with the safety materials, such as fire extinguishers, fire blankets and eye wash, etc., and their location.
4. All hazardous reagents should be used in a well-ventilated hood.
5. Any flammable reagent should be kept away from motors, ovens, or hot areas.
6. Waste hazardous reagents should never be poured into a sink. They must be collected in a waste can.
7. Reaction vessels should be protected by wrapping in metal screen or wire mesh.

■ 4.4 Edit this page from a glossary prepared to accompany a public report on breast cancer. ■

Cyst. a closed cavity or sac which contains liquid or semi-solid material.

Cystitis. Inflammation of the urinary bladder.

Cytology. The study of the original beginning, structure, and functions of cells.

Diagnosis. The definitive determination of the presence and nature of a disease

Edema. The presence of abnormally large amounts of fluid in the body characterized by swelling or puffiness

Estrogen Receptor Assay. A diagnostic test that determines whether or not a cancer's growth is dependent on estrogen.

Excisional Biopsy. Total surgical removal of tissue to be examined

Hematuria. Hematuria is blood in the urine

Incidence. The rate at which a certain event occurs, as in the amount of new cases of cancer occurring during a certain period of time. This is in contrast to prevalence, which refers to the total amount of cases in a population.

In-Situ Cancer. Cancer confined to the site of origin without invasion of local neighboring tissues.

■ 4.5 Edit for grammatical errors. ■

What Is Radon

Radon is a radioactive gas which occurs in nature. It cannot be seen or smelled and you can't taste it.

Where Does Radon Come From?

Radon comes from the natural breakdown (radioactive decay of uranium. Radon can be found in high concentrations in soils and rocks containing: uranium, granite, shale, phosphate, and pitchblende. Radon may also be found in soils contaminated with certain types of industrial wastes, such as byproducts from uranium or phosphate mining, etc.

In outdoor air, radon is diluted to such low concentrations that it is usually harmless, however once inside an enclosed space such as a home radon can accumulate. Indoor levels depend both on a buildings construction and on the concentration of radon in the underlying soil.

How Does Radon Effect Me?

The only known health effect associated with exposure to elevated levels of radon are an increased risk of developing lung cancer. The time period between exposure and the onset of the disease may be many years.

Scientists estimate as many as about 20,000 lung cancer deaths a year in the United States may be attributed to radon.

The risk of developing lung cancer from exposure to radon depends upon the concentration of radon and the length of time you are exposed. Exposure to a slightly elevated radon level for a long time may present a greater risk than exposure to a significantly elevated level for a short time. In general, your risk increases as the level of radon and the length of exposure increases.

■ 4.6 Edit for grammatical errors. ■

A new computer guided robot that can detect cracks in the walls of nuclear reactors has been fabricated by engineers at the Smith River nuclear weapons plant. Under development for two years the $20 million robot is central to efforts to repair the plant's reactors. All five of the reactors are shut down because of serious flaws.

The structural flaws in the reactors developed from an unanticipated weakness in the stainless steel used to build various different components. Containing 19 percent chromium, 8 percent nickel, and less than .08 percent carbon, engineers thought the steel would be impervious to heat generated by the reactors, preventing the development of cracks along the boundaries of the crystalline grains that form steel.

In the year 1980, however, engineers discovered that a stainless steel nozzle that drained one of the plant's reactors had developed a leak. They attempted to repair the crack by breaking through the five feet of concrete that surround the reactor and welding a patch onto the pipe. The repair failed within two years.

During the next four years' time, 31 other leaks developed in pipes, all the result of dete-

rioration in the outer walls of the crystalline grains. To understand the phenomenon, an intensive program was begun by researchers.

Some engineers fear the robot may discover cracks so large that they cannot be repaired which would mean the reactors would have to be shut down permanently.

The new robot will be used to inspect the welds of the reactors. A difficult maneuver, the only openings to the reactors interior are the holes, four inches long in diameter, in the lids. Fuel rods are loaded through these hole shapes.

Engineers with Jared & Kahn the plant's manager for 39 years have long used robots designed to fit through the holes to inspect and repair plant equipment. The challenge was to adapt the robot so that more complex equipment could be housed in the narrow, 20 foot long aluminum rods.

Virtually every aspect of the plant's operations are being overhauled to improve safety.

The Smith River reactors are the nation's only source of tritium a crucial component of fuel for nuclear weapons. Tritium a radioactive isotope of hydrogen decays steadily and must be replenished periodically in most thermonuclear warheads.

■ 4.7 Edit for errors in grammar, usage, and punctuation. ■

Almost everyone in the U.S. gets married at some time in his life. The United States has long had one of the highest marriage rates in the world, and even in recent years it has maintained a relatively high rate. For the cohort born in 1945, for example, 95 percent of the men have married, compared to 75% in Sweden. The other countries studied ranked somewhere between these two extremes.

According to Table 2, a trend toward less marriages is plain in all of the countries studied, although the timing of this decline differs from country to country. In Scandinavia and Germany, for example, the downward trend in the marriage rate was already evident in the 1960s; in the United States, Canada, Japan, France and Holland and the U.K. The decline began in the 1970s.

In Europe the average age at marriage fell until the beginning of the 1970s, when a complete reversal occured. Postponement of marriage by the young is now common throughout the continent. The generation born in the early 1950's initiated this new behavior, characterized by both later and less frequent marriage. Average age at first marriage has also been rising in the United States since the mid-1950s, but Americans still tend to marry earlier than their european counterparts. For example, the average age at first marriage for American men and women in 1988 was 25.9 and 23.6. In Denmark it was 29.2 for men and 26.5 for women.

The high U.S. marriage rate is in part, related to the fact that the United States has maintained a fairly low level of nonmarital cohabitation. In Europe—particularly in Scandinavia, but also in France, the United Kingdom, and the Netherlands there have been large increases in the incidence of unmarried couples living together. This situation is reflected in the lower marriage rates of these countries. Swedish data that includes all cohabiting couples indicates that family formation rates have remained stable since 1960, even though marriage rates have dropped.

Divorce rates have shown a long-term increase in most industrial nations since around the turn of the century. After accelerating during the 1970s, the rates reached in the 1980s are probably the highest in the modern history of these nations. While a very large proportion of Americans marry, their marital breakup rate is by far the highest among the developed countries. Based on recent divorce rates, the chances of a first American marriage ending in divorce are today about one in two, the corresponding ratio in Europe is about one in three to one in four.

Liberalization of divorce laws came to the United States well before it occurred in Europe, but such laws were loosened in most European countries beginning in the 1970s, with the further liberalization taking place in the 1980s. Consequently, divorce rates are rising rapidly in many European countries. By 1986, the rate had quadrupled in the Netherlands and almost tripled in France over the levels recorded in 1960. The sharpest increase occurred in the United Kingdom, where the martial breakup rate increased sixfold. Although divorce rates continued to rise in Europe in the 1980s, the increase in the United States abated,

and the rate in 1986 was slightly below that recorded in 1980. In Canada, although divorce rates remain considerably lower than in the United States, the magnitude of the increase sine 1960 has been greater than in the United Kingdom.

Italy is the only European country studied in which the divorce rate remains low, and divorce laws have not been liberalized there. Japan's divorce rates are lower than in all other countries except Italy, but until Italy, there has been an upward trend in Japan since 1960.

Divorce rates understate the extent of family breakup in all countries: marital separations are not covered by the divorce statistics, and these statistics also do not capture the break-up of families in which the couple is not legally married. Studies show that in Sweden, the breakup rate of couples in consensual unions is three times the dissolution rate of married couples.

■ 4.8 Edit this instructional sheet provided by a government regulatory agency for style. ■

HPLC System Suitability Tests for Potential Impurities

Concerning the submission of analytical samples for methods validation there is the question of what procedures are currently available for the FDA analyst to use to gain some experience with the determination of impurities. For example, it is of interest to FDA to determine to what extent the various impurities may be chromatographically resolved from each other during the analytical procedure. It would be convenient if the applicant provided FDA with procedures for being able to detect and quantitate each of the various impurities potentially present in the sample, whenever possible. This would allow FDA to have the assurance that the analytical method used to monitor the impurities is workable. The concern here is that FDA should have the opportunity to investigate methodology independent of how much of a particular amount of an impurity may exist in a specific lot or sample used in the test. FDA should have available certain testing procedures that may be apart from the applicant's current quality control release methodologies. These could be considered as system suitability tests that focus on the ability of the analytical method to perform its intended separations and quantitations completely independently from any knowledge of whether or not any specific sample contains an amount of an impurity that is analyzable or not. For example, in a worst-possible case scenario, it can be envisioned that if a sample of an applicant's drug was extremely purified so it only had very small amounts of impurities, that is, below threshold detectability limits—then the FDA analyst would have no satisfactory methodology to determine if the proposed analytical method could actually resolve these subject impurities from each other. There is also problematic the additional question of whether an analytical method has the ability to actually detect an impurity at levels which are specified in the control specification. For example, in the event the level of an impurity falls below its threshold detection limit there is the uncertainty for the the FDA analyst to consider that column-to-column variability might not be the factor responsible for the absence of the appearance of a particular impurity peak. From these above considerations it can be seen that it is necessary for the applicant to develop procedures which will allow FDA analysts to have a confidence that any analytical results obtained on an actual test sample of the drug are independently valid—apart from any impurity-to-impurity variability that may be associated with any particular lot number of the drug involved. This approach would mean that the applicant should identify which impurities may be considered as the main ones and which ones can be isolated and structurally characterized as compared to those that cannot be so defined. Impurities not provided in these known manners should be described as accurately as possible in other terms (e.g., lot-to-lot frequency of appearance, quantitative variability plans to more fully analytically investigate them in future designed studies).

Threshold detection limits should be experimentally determined for an impurity, whenever possible; this will help to allow for comparisons to be made in terms of column-to-column variability—this is important if FDA must use an alternate column that is readily available for the validation work. The instructions in these developed analytical procedures (i.e., used as system suitability tests by FDA analysts) should include both procedures for sample preparation of each impurity ion the HPLC column. To be most relevant for comparative purposes, loadings should be proportional to amounts actually achieved in practice on a lot-to-lot basis.

To achieve a maximum degree of accuracy for quantitation, responses factors should be determined—whenever applicably or experimentally possible—relative to the response of that of the active drug being analyzed. This latter adjustment is needed to avoid any ambiguity that is related to the expression of the presence of the amount of an impurity (e.g., often such amounts are mistakenly given in terms of the % total—without any adjustment for response factor differences). In the event that differences in such response factors may require adjustments to be made in the calculation blocks (i.e., correction factor(s) to be used), then this should also be effected.

As with any validation package, reference samples of impurities should be provided in a manner that can be conveniently accounted for by FDA. For example, containers should be identified by chemical name, lot number, and amount present as appropriate. Enough material should be included to permit the designated tests to be conducted. Any precaution for storage and handling should be provided. The actual results of any system suitability testing conducted by the applicant on the lots provided should be included in the validation package whenever possible to permit a close focus for comparisons between laboratories involved.

There should be provided adequate reference to the system suitability tests for the particular HPLC column(s) involved, if not appropriately referenced in the HPLC methodology that is provided in the applicant's quality control release protocol.

It would also be helpful to have available information concerning the linearity of the response of each impurity over its normal range of quantitation if such information is possible since this supports the contention that there is a valid correlation between amounts present and signal response. This would help to show that the applicant has given adequate attention to the matter of being able to define the extent of or lack of quantitativeness (i.e., semiquantitativeness) in the region of analytical interest.

5/////

EDIT FOR TABLES THAT DISPLAY DATA VIVIDLY AND CONCISELY

■ 5.1 Edit this table, prepared to accompany a report showing a greatly improved outlook for college graduates, 1988–2000, so that the title is both informative and visually distinct. ■

Table 1.

	High school or less	1 to 3 years of college	4 or more years of college
Total	100	100	100
Managerial, professional and technical occupations	10	29	68
Sales, administrative support, and service occupations	47	49	25
Craft, operative, laborer, and agricultural occupations	43	22	7

■ 5.2 Edit this table. ■

Table 1. Comparison of Effectiveness of Calcium Polycarbophil vs. Kaolin-Pectin in Treating 224 cases of Acute Diarrhea in Children

| | Results | | | |
	Excellent	Good	Fair	Poor
Calcium Polycarbophil				
Very severe (67 cases)	30	17	14	6
Severe (29 cases)	11	8	73	
Moderate (20 cases)	7	9	40	
Mild (11 cases)				
Kaolin Pectin				
Very severe (56 cases)	8	14	17	17
Severe (15 cases)	1	7	52	
Moderate (16 cases)	6	4	42	
Mild (10 cases)	3	2	1	4

■ 5.3 Edit this table. ■

Composition and Value of Foods

Percent by weight: (Protein, Fat, Carbohydrate)   Mineral Content

| | | | | | | Phosphorus, Iron, Calcium |
			Percent by weight:			Cheese
28.8	35.9	.3	.93	.68	.0013	American
21	22		.92	.68	.0013	Camembert
27.7	36.8	4.1	.92	.68	.0013	Cheddar
25.9	33.7	2.4	.92	.68	.0013	Cream
13.4	10.5		.067	.18	.0030	Eggs Plain
12	18		.087	.24	.0041	Scrambled
21.8	12.1		.024	.25	.0012	Fish: Salmon, Canned
13.5	8.1		.024	.25	.0012	Fresh Salmon
23.8	20.0	.6	.028	.29	.0014	Tuna, Canned, Oil
17	3		.019	.20	.00090	Tuna, Fresh

Note: Proportions of protein, fat, carbohydrate and mineral matter are given as percent by weight of the food named.
Source: U.S. Department of Agriculture

■ 5.4 This table of contents is from a manual for a copy machine. Edit it, using devices such as type appearance (boldface, capitals) and space (indentation, bullets), so the reader can easily see the divisions of main points/supporting points in the manual. ■

Contents

Switching to the OT400
An introduction to the OT400
Identifying your version of the OT400. (locating parts of the OT400; locating parts of the document handler.)
The control panel.
Making copies (using the document handler; using the document feeder; using the document glass)
2-sided copying
Reduction and enlargement
Using pre-set reduction and enlargement
Using variable reduction and enlargement (Using the reduction/enlargement grid Using the reduction/enlargement formula)
Copying Standard Computer forms
Computer form misfeeds
Adding dry ink
Adding Staples
Loading the paper trays (adding paper to the

main paper tray; adding paper to Paper Tray 2; Adding paper using the Paper Tray Bypass)
Ordering supplies
Problem solving

■ 5.5 Here is an entry from a troubleshooting (problem/possible cause/possible remedy) chart. Users with a problem look in the chart, which refers them to the sections and pages in the manual they will need.

The chart is poorly designed. Improve it. Use the stub column for problems. Be sure to label columns. Indent when necessary to show subentries. ■

Problem: Engine will not start
Possible Cause: Carburetor problems: choke operation; needle valve sticking; fuel cut solenoid valve not open.
Remedy: For choke operation, check choke system (see pp. AR 47–53.) For needle valve, check float and needle valve (see p. AZ-12); for fuel cut solenoid valve not open, check fuel cut solenoid valve (pp. FU-12, 13).

Problem: Engine hesitates
Possible Cause: Carburetor problems: float level too low; power valve faulty; choke valve stuck open; fuel line clogged.
Remedy: For float level, adjust float level (p. FU 15). For power valve, check power piston and valve (p. FU 12). For choke valve stuck, check choke system (see pp. EC-47-53). For

fuel line clogged, check fuel line (see pp.G-47-49).

Problem: Insufficient fuel supply to carburetor
Possible Cause: Fuel filter clogged; fuel pump faulty; fuel line clogged; fuel line bent or kinked.
Remedy: (listed respectively for causes) Replace fuel filter; replace fuel pump; check fuel line; replace fuel line (all in FU-31).

■ 5.6 Edit this problem/solution chart. It is part of a manual for a copying machine. ■

Problem Solving

1. If there are spots or marks on copies, there may be similar marks on originals. Try office correction fluid to clean up the originals. The problem may arise from very thin originals. Try using a white backing sheet. The glass may be dirty.

2. If the copies are blank, the originals may be loaded incorrectly. Check to make sure that you have placed them in the document handler *face up*. If you are using the document glass, originals should be *face down*.

3. If the stapler keeps jamming, you will need to clear the jam. To do this, open the finisher door, lift the blue catch and pull out the staple cartridge; remove the staple remover tool from inside the door. Using the tool, remove the jammed staple from the head of the staple cartridge.

■ 5.7 Edit the columns in this table for alignment. ■

Table 11.3. The Solubilities of Some Substances

| Compound | Solubility, g solute/100 g solvent, in water at | | Other solvents |
	0°C	100°C or as specified	
NH_3	89.5	7.4	Organic solvents
NH_4NO_3	118	871	Alcohol, ammonia
$CaCl_2$	59.5	159	Alcohol
$CuSO_4 \cdot 5H_2O$	31.6	203.3	
HCl	82.3	56.1 at 60°C	Alcohol, benzene
MgO	6×10^{-4}	8×10^{-3} at 30°C	
AgF	182	205	
AgCl	7×10^{-5}	2×10^{-3}	

6/////

EDIT FOR FIGURES THAT IDENTIFY AND EXPLAIN

■ 6.1 Below is a page from a home repair manual. Edit to improve the identification and explanation of the figures. ■

Repairing Cracks in Concrete Sidewalks

What You Need:

- packaged, ready-mixed mortar
- bucket
- epoxy concrete ("clear" type for narrow cracks or "gray" type for wide cracks)
- chisel
- wire brush
- pointed trowel and wood float
- paint thinner

How To Do It:

Caution: Repair Only When Concrete Is Dry

1. Chisel out the crack, widening it under the surface.
2. Clean the concrete surface thoroughly with a wire brush.
3. Mix a batch of mortar according to the directions on the package. Mix in the epoxy concrete with the mortar according to the directions on the epoxy container.
4. Put mixture into crack, using the trowel.

Note: Work quickly. Most epoxies harden within an hour. Should the patch harden before you are finished, apply a second coat and smooth the surface again.

5. Using a wood float, smooth the mixture even with the concrete surface.
6. Clean the tools immediately with paint thinner. [19]

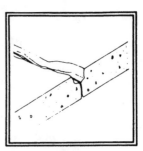

Fig. E.4.

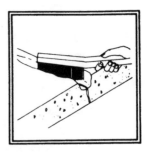

Fig. E.5.

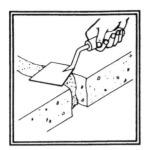

Fig. E.6.

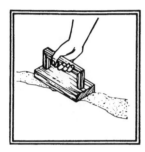

Fig. E.7. *Source:* U.S. Department of Agriculture.

■ 6.2 Below is a page from a home repair manual. Edit for figure identification and explanation. ■

How To Fix Problem Doors

What You Need:

- oil
- graphite
- screwdriver
- hammer
- sandpaper
- plane

- drop cloth

How To Do It:

For Noise

1. You can usually stop a door squeak by putting a few drops of oil at the top of each hinge. Move the door back and forth to work the oil into the hinge. If the squeaking does not stop, raise the pin and add more oil.
2. Use graphite to reduce noise in locks.
3. To stop the rattle in the knob, loosen the set-screw on the knob. Remove the knob. Put a small piece of putty or modeling clay in the knob. Put the knob back on. Push it on as far as possible. Tighten the screw.

For Sticking or Dragging Doors

1. Tighten screws in the hinges. If screws do not hold, replace each screw, one at a time, with a longer screw. As an alternative, insert a matchstick in the hole and put the old screw back in.
2. Look for a shiny spot on the door where it sticks. Open and close the door slowly to find the spot. Sand down the shiny spot. Do not sand too much, or the door will not fit as tightly as it should.
3. If the door or frame is badly out of shape, you may have to remove the door and plane the part that drags.[20]

Fig. E.8.

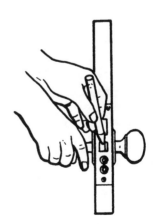

Fig. E.9.

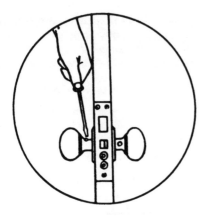

Fig. E.10.

Fig. E.13.

Fig. E.11.

Fig. E.14. *Source:* U.S. Department of Agriculture.

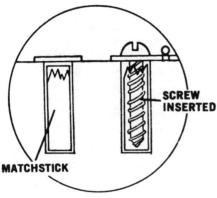

SCREW INSERTED

MATCHSTICK

Fig. E.12.

■ 6.3 The following selections are from a booklet describing methods to lower radon concentrations. The booklet is written for homeowners who have had their homes tested for radon and have decided that they need to take some action to reduce radon levels.

Assume that the text is accurate, and in the correct order. Edit for effective figure identification and explanation. ■

Natural Ventilation

Natural ventilation is an effective, universally applicable radon reduction technique. If done properly, natural ventilation is consistently capable of high reductions of about 90 percent if a sufficient number of windows and vents are opened. High reductions result because natural ventilation both reduces the flow of soil gas into the house by neutralizing the pressure difference between indoors and out, and dilutes any radon in the indoor air with outdoor air.

Natural ventilation, done in the simplest way by opening windows when weather permits, will replace radon-laden indoor air with outdoor air and neutralize pressure. Windows are not the only outlet. Some natural ventilation occurs in every house as air is drawn through tiny cracks and opening by temperature and pressure differences between indoor and outdoor air. In the average American house, outside air equal in volume to the inside air infiltrates about once every hour. In technical terms, this is called 1.0 ach (air changes per hour). Newer houses, which are generally tighter, may have air exchange rates as low as 0.1 ach (one-tenth that of the average house). The rate in older houses, on the other hand, may be more than twice the average (2.0 ach).

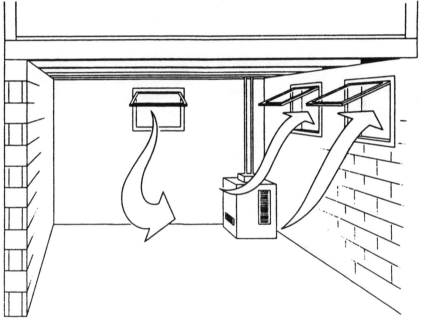

Fig. E.15.

Limitations. The primary shortcoming of natural ventilation is that extreme weather makes the technique impractical year-round in most parts of the country, due to discomfort and increased heating and cooling costs. Open windows can also compromise the security of the house.

Cost. There are no installation costs unless devices must be purchased to hold windows or vents in an open position, or to detect or prevent unauthorized entry through these openings.

Use of natural ventilation in cold weather will increase your heating costs substantially. For example, if you were to increase the air exchange rate to eight times its normal level in your basement and still maintain comfortable temperatures there, your annual house heating bill could be as much as three times greater than normal.

If you normally run an air conditioner in hot weather, your cooling costs will also be greater.

Process. Radon is drawn into your house when the air pressure in the basement or lowest level is less than the air pressure in the surrounding soil. (The primary source of radon is soil.) It is imperative that any ventilation system does not further reduce the air pressure within the house and increase the "pull" that draws radon in. To guard against this, be certain to open vents or windows equally on all sides of the house. Also avoid the use of exhaust fans.

If you ventilate your basement, you may find it more economical or comfortable to close it off and limit its use.

When ventilating unheated areas, be sure to take precautions to prevent pipes from freezing.

Forced Ventilation

Forced ventilation replaces radon-laden indoor air with outdoor air and neutralizes pressure if the fan is big enough. It uses fans to maintain a desired air exchange rate independent of weather.

Rather than relying on natural air movement, forced air fans can be used to provide a controlled amount of ventilation. For example, a fan could be installed to blow fresh air continuously into the house through the existing central forced-air heating ducting and supply registers. Windows and doors would remain closed. Alternatively, fans could blow air into the house through protected intakes through the sides of the house, or could be mounted in windows. A fan could also be installed to blow outdoor air into a crawl space.

Cost. The installation costs for forced-air systems ranging from simple window fans to elaborate heating, ventilation, and air conditioning systems will vary from $25 to $1,000. The additional cost of electricity for forced-air systems will vary depending upon the size of the fans, the number of fans used, and the amount of use. A single window fan can have electricity costs as low as $20 per year, while a central furnace fan may cost $275 a year to operate.

Use of forced ventilation during cold weather will substantially increase heating costs. As with natural ventilation, if you were to increase the air exchange rate to eight times its normal level in your basement while maintaining comfortable temperatures there, your annual house heating bill could be as much as three times greater than normal.

Reductions. Tight houses with low exchange rates are likely to benefit more than are houses with high exchange rates. In a typical house, to achieve a 90 percent reduction of radon, you will probably need a 500 to 1,000 cfm (cubic feet per minute) fan.

Limitations. Forced ventilation, like natural ventilation, can be employed in most houses, but in many cases the trade off in decreased comfort and/or excessive heating or cooling costs may prove unacceptable. This approach may be useful as an interim measure with very high radon levels.

Procedure. You should ventilate the lowest level of your house. Closing off and not using a basement may also be advisable. Air should be blown **into** the house and allowed to exit through windows or vents on adjacent or opposite sides. In many homes, blowing air in through an existing central furnace is quite practical. **The use of an exhaust fan to pull air out of the house may decrease the interior air pressure and draw more radon inside.** The use of whole-house fans is not recommended because they typically operate in the exhaust mode.

Air distribution and ventilation rates can be controlled by the sizing and location of fans and the use of louvered air deflectors. When ventilating unheated areas, be sure to take precautions to prevent pipes from freezing.

Sealing Cracks and Openings

Radon is a gas that can pass through any opening in a floor or wall that touches the soil. It can enter your house through

- openings around utility pipes
- joints between basement floors and walls, including perimeter drains
- other floor drains (especially those that discharge into dry wells)
- the holes in the top row of concrete blocks
- tiny cracks and openings such as the pores in concrete blocks

Sealing such cracks and openings is often an important preliminary step when other methods

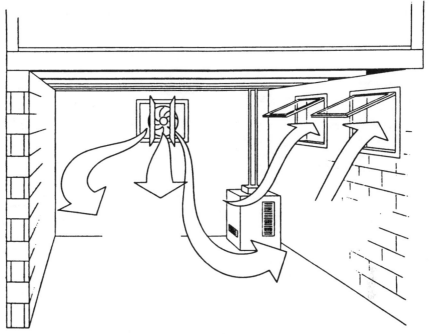

Fig. E.16.

are used. For houses with marginal radon problems, sealing alone may be sufficient.

In some houses, certain areas will be difficult, if not impossible, to seal without significant expense. These include the top of block walls, the space between block walls and exterior brick veneer, and openings concealed by masonry fireplaces and chimneys.

Installation. Since effective sealing generally requires meticulous surface preparation and carefully controlled application of appropriate substances, the work is often most effectively done by experienced and competent contractors or highly skilled homeowners.

Costs. The costs are highly variable. Do-it-yourself closure of accessible major entry points can be inexpensive. Putting traps in drains and covering sumps can be low to moderate in cost. Applying membranes and coatings can be expensive.

Reduction. If sealing is done thoroughly, and all exposed earth is covered, reductions may be

sufficient in some houses. In others, when sealing is used alone, you should expect only low to moderate reductions.

Limitations. It is very difficult to find all the cracks and openings in your house. This method may have little effect on radon entry unless nearly all the entry points are sealed. Furthermore, settling of the house and other stresses may create more cracks as time passes. Also, the openings in the top row of concrete blocks in a wall are often inaccessible or otherwise difficult to seal tightly. As a house settles and reacts to external and internal stresses, old seals can deteriorate and new cracks can appear. The aging process ultimately ends the ability of sealants to block out soil gases. Therefore checking and maintenance are required at least yearly.

Procedure. Holes in the top row of concrete blocks in basement walls should be sealed with mortar or urethane foam.

Seal wall and floor joints with flexible polyurethane membrane sealants. Cracks and utility openings should be enlarged enough to allow filling with compatible, gas-proof, non-shrinking sealants.

A water trap should be installed in floor drains connecting to drainage or weeping-tile systems. Water traps allow water that collects on basement floors to drain away but greatly reduce or entirely eliminate entry of soil gas, including radon. Water traps must be kept filled with water to be effective.

Perimeter drains should be filled with urethane foam; however, some alternative plan for water drainage should be provided.

Porous walls, especially block walls, require the application of waterproof paint, cement, or epoxy to a carefully prepared surface.

Heat-Recovery Ventilation (HRV)

This method replaces radon-laden indoor air with outdoor air.

A device called a "heat recovery ventilator" uses the heat in the air being exhausted to warm the incoming air. In an air-conditioned house in warm weather, the process is reversed: The air being exhausted is used to cool the incoming air. This saves between 50 and 80 percent of the warmth (or coolness) that would be lost in an equivalent ventilation system without the device.

Installation. Ducted units are designed, installed, and balanced by experienced heating/ventilation/air conditioning contractors. Wall-mounted units are generally less complex, and can sometimes be installed directly by the homeowner.

Cost. Installation costs (materials and labor) will range from $800 to $2,500 for ducted units. Wall-mounted units cost about $400.

The cost for electricity to operate one of the larger units with two 200-cfm fans is about $30 per year.

Using a heat recovery ventilator could save you 50 to 80 percent of the increase in heating and cooling costs that would result from achieving a comparable amount of ventilation without heat recovery.

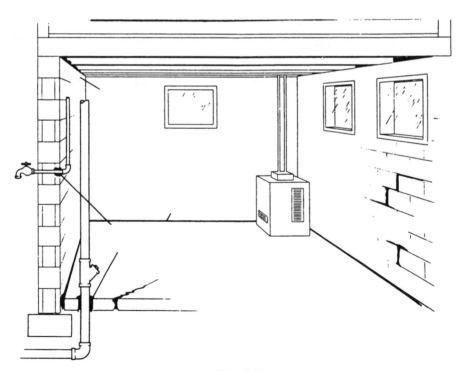

Fig. E.17.

Reductions. A radon reduction of 50 to 75 percent can be achieved in houses of typical size and infiltration rate, assuming between 200 and 400 cfm of HRV capacity. Reductions can be greater in tight houses. Reductions will vary throughout the house, depending on ducting configurations.

Limitations. The applicability of HRVs for radon reduction will likely be limited to situations where only moderate reductions are needed and where winters are cold. If an HRV is intended to serve as a stand-alone measure to achieve 4 picocuries per liter (pCi/L) in a house of typical size and infiltration rate, the initial radon in the house could be no greater than 10 to 15 pCi/L. Greater reductions can be achieved in tight houses.

Procedure. To simplify the necessary ducting runs to different parts of the house, the heat recovery ventilator unit, consisting of the core

and fans, can be located in an inconspicuous part of the house, such as an unfinished basement or utility room. Care must be taken to keep fresh air supply registers well-removed from return air withdrawal points. Locate the radon-laden air returns in the basement or lowest level. It is crucial that the flow-rates in the fresh air intake duct and the radon-laden air exhaust duct be balanced. If more air is exhausted than is brought in, the house will become depressurized and even more radon may be drawn into the house. Be sure the balancing is done with no pressure difference between indoors and outdoors, since the unit will tend to maintain any pressure difference that exists when it is balanced.

Heat recovery ventilators are usually cost-effective only if operated during cold weather or in hot weather if the indoor versus outdoor temperature difference is large. At other times, the same amount of ventilation and radon removal can be achieved by simply opening windows.[21]

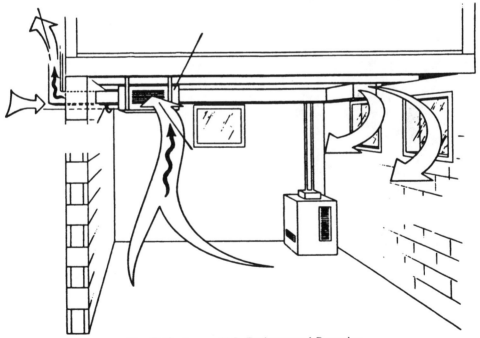

Fig. E.18. *Source:* U.S. Environmental Protection Agency.

7/////

EDIT FOR COMPLETENESS, ACCURACY, AND CONSISTENCY

■ 7.1 Edit the following passage from a technical article for completeness and accuracy. ■

Table 7. Percent of French Men and Women in Marriages or Consensual Unions, by Age, 1988

Sex and Age	Married	In Consensual Union
Men:		
18–24	4.7	6.1
18–19	0	.1
20–24	6.5	8.4
25–29	42.7	14.5
30–34	67.4	9.8
35 and over	78.7	3.4
Women:		
18–24	14.0	10.4
18–19	.7	1.8
20–24	19.0	13.7
25–29	55.9	12.3
30–34	71.7	7.6
35 and over	63.5	2.1

Source: Institut National de la Statistique et des Études Économiques, *Enquête sur l'emploi de 1988: résultats détaillés* [Labor Force Survey of 1988: Detailed Results], Les Collections de L' INSEE, Série D, no. 128 (Paris, INSEE, October 1988), table MEN-07, pp. 104–05.

In France, nonmarital cohabitation increased from 3 percent of all couples in 1975 to more than 6 percent in 1982 and 8 percent in 1988. Table 3, which shows the percent of all French men and women in consensual unions or marriages by age group in 1988, illustrates the fact that cohabitation occurs predominantly in the younger age groups.

As in France, the younger age groups in Sweden have a higher incidence of cohabitation. For instance, in 1980, 4 out of every 5 unmarried Swedish men aged 20 to 24 were living in a consensual union, as were 68 percent of all unmarried women in that age group. In the age group 25 to 29, the proportions were 49 percent and 35 percent, respectively. Virtually all Swedes now cohabit before marriage.

Table 9 focuses on participation rates of

Table 9. Labor Force Participation Rates of All Women under Age 60[1] and Women with Children under the Ages of 18 and 3, Eight Countries, 1986 or 1988[2] [In percent]

Country	All Women	All Women with Children		Lone Mothers[3] with children	
		Under 18 Years Old	Under 3 Years Old	Under 18 Years Old	Under 3 Years Old
United States	68.5	65.0	52.5	65.3	45.1
Canada	66.8	[4]67.0	58.4	[4]63.6	41.3
Denmark	79.2	86.1	83.9	85.9	80.9
Germany	55.8	48.4	39.7	69.7	50.4
France	60.1	65.8	60.1	85.2	69.6
Italy	43.3	43.9	45.0	67.2	68.0
Sweden	80.0	[4]89.4	[5]85.8	([6])	([6])
United Kingdom	64.3	58.7	36.9	51.9	23.4

[1]Women ages 60 to 64 are included in Canada and Sweden. Lower age limits are 16 for the United States and Sweden, 15 for Canada, and 14 for all other countries. For participation rates of women with children, no upper limit is applied for the United States or Canada. Theses differences do not distort the comparisons because very few women under 16 have children, while few women over 60 live with their children.
[2]Data for the United States are for March 1988; Canada and Sweden—annual averages for 1988; data for all other countries are for spring 1986.
[3]Includes divorced, separated, never-married, and widowed women.
[4]Children under 16 years.
[5]Children under 7 years.
[6]Not available.
Sources: Published data from U.S., Canadian, and Swedish labor force surveys; unpublished data for other countries provided by the Statistical Office of the European Communities from the European Community labor force surveys.

women with children under the age of 18 and under the age of 2 in a recent year in seven countries. Except for Italy, women with younger children tended to have lower participation rates than women with children under age 18. Danish and Swedish women continue to stand out, with more than 8 out of every 100 women with younger children participating in the work force. . . . French and Canadian women, with 6 out of 10 economically active, were second to the Scandinavian women. In the United States, about 5 out of 10 women with children under age 3 were in the labor force. The participation rates for German and British women were substantially lower than in other countries.

Although no historical data are shown in table 9, it is clear that there has been a dramatic increase in participation rates of women with younger children. For example, about 40 percent of Swedish women with children under the age of 7 (the age at which compulsory schooling begins) were employed in 1970; today, 85 percent are working. In Canada, women's overall participation rate increased from 45 percent in 1976 to 55 percent in 1986, and the greatest increase involved women with children under 3 years of age.

Table 9 also shows participation rater for mothers without partners. In the United States, Canada, Denmark, and Britain, single mothers with young children had lower participation rates than all mothers with young children. By contrast, in France, Germany, and Italy, single mothers of young children had higher participation rates than their married counterparts.[22]

8/////

FINAL EXERCISE: EDIT FOR A COMBINATION OF ERRORS

Assume that the following two chapters were written for inclusion in a booklet on drying eastern hardwood lumber. The prospective audience includes (1) people interested in drying small quantities of lumber inexpensively, (2) people responsible for hardwood lumber processing in mills, custom drying operations, and furniture plants, and (3) teachers and students in related career development programs.

The document needs editing. Some of the problems that might be addressed include these:

1. Lack of prefatory summaries or abstracts before chapters and some subsections.
2. Poor use of headings and subheadings.
3. Erratic use of definitions
4. Multiple errors in style, grammar, and usage.
5. Displays in tables.
6. Identification in figures.
7. Inconsistencies between text, references, and figures. ■

Chapter 2   Special Predrying Treatments

Using Steaming to Accelerate Drying, Modify Color, and Recover Collapsed Wood

Accelerating Drying. Early in the study of wood drying at the U.S. Forest Products Labo-ratory observations were made that steaming hardwoods before drying sometimes results in reduction in drying time. Detrimental affects were noted however, when air dried stock containing surface checks was steamed. The checks deepened and widened. The steaming used at that time was relatively long. The conclusion was drawn that such steaming represented a delay during which no drying occurred and general use of the practice was stopped.

More recent studies in a number of schools and laboratories have shown a number of benefits. The permeability of both softwoods and hardwoods are increased by short periods of steaming. Moisture migration rates are increased significantly, and drying times are reduced. The development of prefabricated aluminum kilns have made possible the use of steaming in commercial drying operations.

Sampson (1975) reviews the most significant research results and investigated the acceleration of drying small specimens of wood. He tried several species of wood in the green condition and steamed at 212 degrees F. Drying rates were increased for northern red oak, cherrybark oak, and sweetgum heartwood. The drying rate at 50 % moisture content for the small specimens increased 34 too 75% for the oaks and 11 to 36% for sweetgum heartwood. Steaming times were in the range of one-half to five hours. For sweetgum heartwood, the 5 hour period was best. The drying rate of sweetgum sapwood was slightly reduced by steaming.

In another study with 1-inch-thick oak, Simpson (1976a) found that the moisture gradients during drying after steaming 4 hours at 212° F, were smooth curves. The natural moisture gradients of the unsteamed controls had inflections that indicated free water movement was restricted. Simpson's work showed that free water migration from the center toward the surface were enhanced by steaming.

The above results were achieved with saturated steam at 212° F, a condition difficult to obtain in commercial kilns. A larger scale study used both green and partly air dried rough northern red oak (Simpson 1976b). The lumber was pretreated with nearly saturated steam for 4 hours. Reductions in drying time for both classes of lumber was about 17 percent. No defects occurred in the green lumber or in one batch of the partly air dried material. The other partly air dried batch, however had been severely surface checked during air drying. The steaming appeared to deepen the surface checks and change them into bottleneck of honeycomb checks. The change to honeycomb checks confirms the admonition not to use steam spray during warmup of a kiln charge of fully air-dried oak. Surface checks may be present in such oak but not visible.

While the above studied are only exploratory, they give some prospect for practically accelerating the drying of eastern hardwoods by presteaming treatments. Additional studies are needed to confirm benefits and determine limitations, comparative energy demands and economics.

To Modify Color

Walnut. The most common use of presteaming treatments is to modify the color of the sapwood of black walnut. Steaming darkens the sapwood, toning down the contrast between it and rich, brown colored heartwood and facilitating the uniform finishing of the wood. It also improves the color of the heartwood, making it and the sapwood more uniform. There is some extraction of coloring matter from the heatwood during the process. These extractives do not penetrate the sapwood beneath the surface.

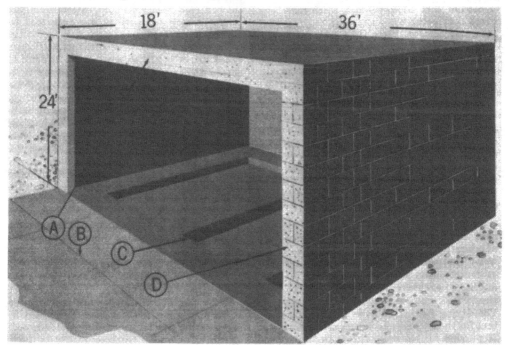

Fig. E.19. Steaming chamber constructed of concrete blocks and poured reinforced concrete. Shown are the roof, poured concrete or prestressed concrete sections; the apron or loading ramp; the trough for perforated or imperforated steam pipes in water; and the walls of concrete blocks or poured concrete.

The best steaming results are achieved by treating green lumber with wet steam in as tight a structure as possible at temperatures which give the most color in the least time.

Nonpressure steaming of walnut is done in special vats or buildings with provisions for wet steam. Any structure is suitable so long as it is made of materials that will stand up under wet heat up to 215°F. Steaming times are 24 to 96 hours. Walnut should not be steamed in the dry kiln because of the amount of time required and the corrosive affects of steam and volatile extractives.

Three steaming vats or chambers are shown in figures 1 to 3. In figure 1, the floor is reinforced concrete and the doors are typical dry-kiln type. The lumber is stacked by forklift trucks on wooden bolsters. Low to moderate pressure steam is introduced by preforated steam pipes in the troughs, which are filled with water. An alternative is to circulate steam in closed pipes submerged in the water in the troughs. The steam traps in this case discharge their condensate into the troughs to keep them full. The capacity of a chamber of the dimensions shown is 40 to 50 thousand board feet of lumber in solid-pile packages.

In Figure 2, the lumber is stacked in the vat in the same manner as in figure 1. The sections of the vat roof are put in place by the forklift as the vat is filled. The front door sections are the same construction as the roof sections. Joints between sections are covered with sawdust to reduce steam looses. The walls of the poured concrete camber in Figure 3 are 10 inches thick. The permanent roof is made of hollow precast concrete sections covered with insulation and built-up roofing. Dry kiln type insulated doors with double surfaces of aluminum are used. Steam from a high-pressure boiler is injected directly in the chamber.

In figure 2, all joints of the flashing-weight aluminum sheets were sealed with a high melt asphalt. Other materials of construction are possible, including solid wood cribbing, aluminum frame and sheathing and tile. All types of steam vats or chambers should be protected by a coating of asphalt or one of the materials supplied by dry kiln companies for kiln coating. Some walnut steam chambers have trowelled mortar-type coatings similar to those used in patching boiler masonry. In any event, iron and steel fittings should not be exposed directly to the steam. Provision should be made

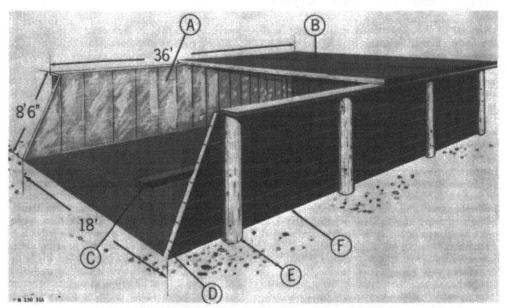

Fig. E.20. Steaming vat of wood and sheet aluminum construction. A, Sheet aluminum lining with joints sealed with asphaltic materials. B is roof sections of creosoted frame and blanks, lined with sheet aluminum. C, slanted jamb causes door section to fit securely by weight alone. D is the trough for perforated steam pipes in water. E, creosoted posts set in ground and F is the creosoted plank wall.

Fig. E.21. Steaming chamber of poured concrete walls and roof. A, drain for condensate. B, steam pipe; C, concrete blocks turned on sides and imbedded in concrete to hold lumber packages off of floor. D is the apron or loading ramp.

for gasketing or sealing doors at tops, sides and bottom.

Walnut steaming chambers preferable are equipped with recording thermometers and some have temperature controllers also. There are no fans in steaming chambers.

Steaming has also been done commercially under pressure. Brauner and Conway (1965) developed the optimum conditions experimentally. Then they settled on steaming at 6 pounds per square inch pressure and 230° F for 5 hours. A longer time is needed in the winter. This procedure not only darkens the sapwood, the heartwood loses its purplish cast and becomes chocolate brown in color. Although the coloration is rapid and time-saving the lumber must be cooled in the pressure vessel or end checking and honeycombing occur. An alternative is to take the load out of the retort and cover the wood with a tarp until cool. Millions of board feet of walnut have been steamed at the Conway plant using a 7- by 8- by 20-foot pressure vessel. At another locations a suitable pressure steamer was made by adapting a preservative treating cylinder door to the end of an old tank car.

Steaming walnut may increase it's drying rate, but there is no conclusive proof of this at this point in time.

Other woods

European beech lumber is frequently steamed to give it a reddish-brown tone which enhances its color when finished appropriately. Beech steaming is not generally done in the U.S. but presumably it could be, if desired. Steam has been used, however, to give oak a brown color. Ordinary kiln-drying conditions can be manipulated to give sugar maple a reddish brown color (McMillen 1976 but presumably the same could be done by a short steaming period followed by customary drying. Sweetgum sapwood was steamed before air drying at one time to sterilize the wood, promote rapid air drying, and avoid blue staining. No record was made of color effects from such steaming.

To Recover Collapsed Wood

Collapse is technically excessive shrinkage and distortion of individual wood cells in various zones of the wood. In a collapse-prone wood the external effect of collapse is a "washboard-type" surface.

In Australia, where collapse prone species are common recovery is accomplished by taking the collapsed boards out of a kiln charge at about 18 percent moisture content, steaming it for 12 to 24 hours in a chamber like that used for walnut, and then finishing the drying in a kiln with noncollapsed wood. Steaming is done with saturated steam at 212°F.

Using Chemical Seasoning

Chemical seasoning is comprised of treating green wood with a hygroscopic chemical and air drying or kiln drying the treated material. The chemical reduces surface checking during seasoning, rather than speed the drying.

The objective is to chemically impregnate the outer zone of lumber with chemicals to a depth of about one-tenth of the thickness, with the highest concentration at or near the surface. The chemicals maintain the outer zone at a high moisture content during early stages of drying. This reduces the shrinkage of the outer zone and lessens the tendency to surface check. Some chemicals additionally impart a certain degree of bulking. Kiln schedules must be modified somewhat to bring the initial relative humidity below the relative humidity in equilibrium with the saturated chemical solution.

Numerous chemicals have been used (MCMillen 1960). Common salt is cheap and effective in reducing surface checking. Urea, which is effective with Douglas fir, was not as effective as salt on oak. Other chemicals included: invert sugar, molasses, diethylene glycol and a urea-formaldehyde mixture. A proprietary salt mixture having corrosion inhibitors in it was popular for a time.

The proprietary chemical as well as common salt, can corrode metals and damage dry-kiln equipment, woodworking machinery, and hardware fastened to the treated wood, respectively, if the amount of treating chemical is excessive or the treatment time too long. Salt-treated wood regardless of care in treatment, will have a corrosive effect on metals in contact with it in regions of prolonged high humidity such as the Gulf of Mexico and South Atlantic coasts, etc. Salt also can reduce the strength of wood and cause problems in gluing and finishing. Although salt has been and continued to be used successfully in the seasoning of thick southern hardwoods, considerable care in treatment, drying, and use is recommended.

Using the Polyethylene Glycol Process

Green wood heavily treated with PEG retains its green dimension during drying and indefinitely, thus the wood is permanently restrained from shrinking, swelling, or warping regardless of atmospheric humidity. For maximum dimensional stability, polyethylene glycol must be diffused deeply into the wood in amounts of 25 to 30 percent of the dry weight of the wood. 2 solutions commonly used are 30 and 50 % PEG by weight. Heating the solution during treatments speeds up the diffusion, but soaking times range from 3 to 30 days. Kiln drying after treatment can in many cases be much more drastic than normal. The process is especially helpful for hardwood tree and limb cross sections, thick novelty items, and carvings and material with highly irregular grain. Details of the process, which is relatively expensive, have been described by Mitchell (1972).

Surface treatments to prevent checking

In addition to presurfacing, there are certain materials which can be applied to the woods surface to retard checking. Chemicals used successfully are: wax, sodium alginate and a salt paste. These materials either retard moisture movement or alter the vapor pressure at the surface. No details are available on use of waxes, but a thick emulsion of microcrystalline wax has been applied to the sides of highly figures gunstocks in California to prevent checking.

Sodium Alginate

In Australia Harrison (1968) investigated the use of very viscous sodium alginate solutions or emulsions as dip treatments on a variety of hardwoods up to 2 inches thick. When the lumber is air dried the alginate dries out to form a porous skin over the surface. It is effective in preventing checking in all of the woods tried under severe air drying conditions. The method has not been tried in the U.S. but might have some benefit for thick oak.

Sodium alginate is a dry powder obtained from seaweed and is used in a variety of products in the United States. Considerable care must be used in mixing it to form the solution. The wood must be still quite green for the treatment to be effective in preventing checking. The air-drying piles must be carefully roofed to keep the alginate from being washed off by the rain.

Salt Paste

The United States Bicentennial celebration inspired widespread interest in seasoning disks or thick sections of large trees. The disks were desired for exhibits or usable items on which the chronology of important events could be shown. It is very difficult to successfully season such disks. The most damaging defect is the large V-shaped check which is likely to develop because tangential shrinkage is usually much greater than radial shrinkage. In addition, many small end checks tend to appear over the entire surface. Polyethylene glycol has promise, but is slow, costly, and the temptation always exists to terminate treatment before all wood cell walls are fully saturated with PEG.

Other Pretreatments

Precompression

Ceck (1971) has established that dynamic transverse compression of 2-inch yellow birch lumber, before drying by a severe schedule, significantly improved drying behavior. The drying was carried out at 215oF. Momentary thickness compression of 7 to 8.5 percent in a roller device had the result of greatly reduced collapse and honeycombing compared with uncompressed material. Drying time was only 8 days compared with a customary 18 days for non-compressed material dried by conventional schedules.

Prefreezing

The freezing of green wood followed by thawing before the start of drying is another treatment which increases the drying rate and decreases shrinkage and seasoning defects in some species. Most research on hardwoods has been done with black walnut (Cooper, Erickson, and Haygreen). The effect on shrinkage appears to be a good indicator of seasoning improvements, and favorable results have been attained with black cherry, American elm, and white oak (Cooper and Barham 1972). Another eastern hardwood species for which decreased shrinkage has been found is black tupelo (blackgum). The best pre-freezing temperature for black walnut is $-100°$ F, but significant improvement is found at $-10°$ F, a temperature readily attained in commercial freezing equipment. A freezing time of 24 hours is adequate for thicknesses up to 3 inches. A similar length of time duration has been used for thawing.

Although prefreezing would require substantial additional investment, a shortening of drying time to half of the time required by some of the industry has been demonstrated for walnut gunstock blanks (Cooper, Bois, and Erickson 1976). Using a slightly accelerated kiln schedule 600 prefrozen gunstock blanks dried from green in 103 days had only 2.66 percent defective from collapse, checking and warp while 6.50 percent of the 400 unfrozen blanks in the pilot test had similar defects.

Thus precompression and prefreezing are two treatments that have some promise of improving the drying rate of hardwoods without decreasing quality.

LITERATURE CITED.

Brauner, A. and E. M. Conway
  1964. Steaming walnut for color. For. Prod. J. 14(11):525–7.
Cech, M. Y.
  1971. Dynamic transverse compression treatment to improve drying behavior of yellow birch. For. Prod. J. 21(2):41–50.
Cech, M. Y. and F. Pfaff
  1975. Kiln drying of 1-inch red oak. For. Prod. J. 25(8):30–37.
Cooper, G. A. and S. H. Barham
  1972. Prefreezing effects on three hardwoods. For. Prod. J. 22(2):24–25.
Cooper, G. A., R. W. Erickson, and J. G. Haygreen
  1970. Drying behavior of prefrozen black walnut. For. Prod. J. 20(1):30–35.
Cooper, G. A., P. J. Bois, and R. W. Erickson
  1976. Progress report on kiln-drying prefrozen

walnut gunstocks—techniques and results. FPRS Separate No. MW–75–S70, For. Prod. Res. Soc., Madison, Wis.

Harrison, J.
1968. Reducing checking in timber by use of alginates. Austral. Timber J. 34(7):24–25.

McMillen, J. M.
1960. Special methods of seasoning wood: Chemical seasoning. U.S. For. Prod. Lab. Rep. 1665-6.

McMillen, J. M.
1976. Control of reddish-brown coloration in drying maple sapwood. USDA For. Serv. Res. Note FPL–0231. For. Prod. Lab., Madison, Wis.

Mitchell, H. L.
1972. How PEG helps the hobbyist who works with wood, 20 p., illus. Pamphlet, USDA For. Serv., For. Prod. Lab., Madison, Wis.

Simpson, W. T.
1975. Effect of steaming on the drying rate of several species of wood. Wood Sci. 7(3):247–255.

Simpson, W. T.
1976a. Effect of presteaming on moisture gradient of northern red oak during drying. Wood Sci. 8(4):272–276.

Simpson, W. T.
1976b. Steaming northern red oak to reduce kiln drying time. For. Prod. J. 26(10):35–36.

CHAPTER 8   OTHER METHODS OF DRYING.

Heated-room drying

In heated room drying a small amount of heat is used to lower the relative humidity. This also lowers the equilibrium moisture content to which the wood will come if left in the room a long time. This method is suitable only for wood that has been air dried first. Green lumber may check and split by this method. This method does not dry lumber rapidly, but it is suitable for small amounts of lumber.

Before air drying is started, the lumber should be cut as close as possible to the size it will have in the product. Allowance must be made for some shrinkage and warping during drying and for a small amount to be removed during planing and machining. If it is necessary to shorten some long pieces of air-dried lumber before heated-room drying was started, the freshly cut ends should be end coated to prevent end checks, splits, and honeycomb.

For reasonably fast heated room drying, the wood should be exposed to an EMC about 2 per cent below the moisture content of use. The wood is left in the room just long enough to come to the desired average moisture content. Then the wood is taken out and stored in a solid pile until it was used. The storage area should have the same EMC as the area in which the wood will be used.

The amount that the temperature must be raised above the average outdoor temperature depends upon the average outdoor relative humidity. Typical values are given in table 1. Do not attempt to use more heating with this method. Any ordinary room or shed can be used and any ordinary means of heating the room should be satisfactory. A slight amount of air circulation is desirable to achieve temperature uniformity. If the material is relatively small in size it can be piled in small, stickered piles on a strong floor. It also could be sticker piled on carts that can be pushed in and out of the room.

Table 1

EMC value temperature at	Degrees above average outdoor desired		
	70 pct RH	75 pct RH	80 pct RH
Percent	Degrees F	Degrees F	Degrees F
4	38	40	42
5	31	33	35
6	23	25	27
7	18	20	23
8	13	15	17
9	10	12	14
10	6	8	10

Long lumber should be piled on strong raised supports as shown in Figure E-22.

Dehumidifier Drying

Considerable interest has been expressed in dehumidifier-type kilns developed in Italy, Germany, Norway, and England. These dryers have been used successfully for hardwoods in Europe where kiln schedules are as a general rule conservative. In these locations the desired final moisture content usually is not below 9 percent. Technical information from the British Forest Products Research Laboratory (Pratt 199968) indicated such dryers were not operated at temperatures over 130° F. Drying of hardwoods to 7 percent moisture content requires use of higher temperatures to be practical.

The dehumidifier dryer uses electrical refrigerator-type equipment in an arrangement favorable to the drying of wood and other materials. In a closed building or room, moist air from the lumber is drawn over refrigerated coils. The air is cooled below it's dew point and part of the moisture is condensed. The water flows to a tray in the base of the unit and is drained out of the kiln. The air circulates past the condenser unit of the system and picks up the latent heat of vaporization. Then the air is blown by fans through the lumber. In some installations, however the dehumidifier is located outside the drying chamber and the latent heat benefit is not realized.

The relative humidity (RH) as a general rule is lower than could be obtained at the same temperature in a conventional kiln vented to the air. Electrical resistance heaters, hot water coils or steam coils can be used to augment the heat from the condenser. Brief coil-heating periods of time can be used to defrost the coils when necessary and interruptions of drying can be inserted in the schedule to minimize drying stresses.

Advantages claimed by manufacturers and borne out to some extent by published data are:

Relatively low capital investment.

Operating costs are not exorbitant.

Drying time to 12 percent moisture content not unduly prolonged for woods normally dried at low temperature.

The equipment lends itself readily to automatic programming so the dryers can be operated by relatively unskilled personnel with little danger of overdrying.

Mixed species and thicknesses can be dried, within certain limitations, in the same dryer.

Low shrinkage and warping.

Many of these advantages are also found in heated-room drying, but heated room drying is best restricted to air-dried material. Dehumidifier dryers can be scheduled to safely dry refractory woods from the green condition.

Solar Drying

The changed world fuel situation in today's modern environment has renewed interest in the use of solar energy to dry lumber. Research in small experimental and semi-commercial solar dryers showed that hardwoods can be satisfactorily dried from the green to the air dried condition. Drying times are roughly about 50 to 75 percent of the air-drying times. Hardwoods can be dried further to 10 percent moisture content, but the time required is long compared to kiln drying.

Peck (1962), using a wood-framed dryer rectangular in shape and with double-layer transparent plastic film walls and roof in Madison, Wisconsin dried northern red oak from 75 to 20 percent moisture content in 23 to 105 days starting in various different months of the year. Drying times from 20 to 10 percent MC were 25 days for a time period starting in July and 42 days starting in the month of September. A 24 inch fan driven by a ⅝-horsepower electric motor circulated the air during daylight hours. The dryer roof, which essentially was the major solar energy collecting surface, had 95 square feet of surface, the south wall had 17 square feet.

Additional research on solar drying of wood has gone on around the world using equipment varying from very small cabinets in India to semicommercial units holding about 3,000 board feet in Uganda. Plumptre (1973) discussed his own research with the Ugandan kilns and reviews world literature on the subject in a United Nations report.

Johnson built a small homemade solar kiln in southwestern Wisconsin to efficiently dry small quantities of hardwood lumber for his own use. Solar heat was collected by single-thickness windows on the south facing wall of an A frame structure. A ventilating slot was built in the wall below the windows. A wind-powered centrifugal blower drew air up over a heat-absorbing metal sheet behind the glass windows and then forced the air down through the lumber edge-piled on a rack. One-inch cherry and white oak were dried to 10% moisture content in 2 to 6 weeks.

Johnson has continued to dry lumber in a dryer of slightly modified design (Boix 1977). The lumber is piled flat. An electric motor is used to drive the fan, but only when the temperature is above 85° F. The air is forced down into a central flue, then horizontally through the lumber pile. He estimated that it takes 400 hours of sunshine to dry hardwood lumber from green to 10 percent moisture content.

While this method of drying is feasible for small quantities of lumber for hobby use, drying times are effected greatly by season of the year and the location. The relative suitability for solar energy collection the year round is only fair in Northeastern United States from Maine to central Wisconsin, including New England, New York, and Pennsylvania, and only good in most of the rest of the Eastern United States. In general it appears that heated-room drying is a more practical way of drying small quantities of hardwoods as long as fuel costs are no more than double what they are now.

Bois, P. J.
    1977. Constructing and operating a small solar-heated lumber dryer. USDA For. Serv. State and Private For., For. Prod. Util. Tech. Rep. No. 7, Madison, Wis.
CEAF
    1976. The drying of solid timber. Seminar for U.N.I.D.O., Milan, Italy, May 18 (Trade brochure, CEAF, Turin, Italy).
Chudnoff, M., E. D. Maldonado, and E. Goytia
    1966. Solar drying of tropical woods. USDA For. Serv. Res. Pap. ITF 2. Inst. Trop. For., Rio Piedras, P.R.
Crowther, R. I., et al.
    1976. Sun, earth: How to use solar and climatic energies today. A. B. Hershfeld Press, Denver, Colo.
Eckelman, C. A. and J. A. Galezewski
    1970. Azeotropic drying of hardwoods under vacuum. For. Prod. J. 20(6):33.
Hittmeier, M. E., G. L. Comstock, and R. A. Hann

1968. Press drying nine species of wood. For. Prod. J. 18(9):91–96.
Huffman, D. R., F. Pfaff, and S. M. Shah
    1972. Azeotropic drying of yellow birch and hard maple lumber. For. Prod. J. 22(8):53–56.
Johnson, C. L.
    1961. Wind-powered solar-heated lumber dryer. South. Lmbrmn. 203(2532):41–2, 44.
McMillen, J. M.
    1961. Special methods of seasoning wood. Boiling in oily liquids. U.S. For. Prod. Lab. Rep. No. 1665 (rev.), For. Prod. Lab., Madison, Wis.
McMillen, J. M.
    1961. Special methods of seasoning wood. Vapor drying. U.S. For. Prod. Lab. Rep. 1665–3 (rev.), For. Prod. Lab., Madison, Wis.
Peck, E. C.
    1962. Drying 4/4 red oak by solar heat. For. Prod. J. 12(3):103–7.
Plumptre, R. A.
    1973. Solar kilns: their suitability for developing countries. U.N.I.D.O. Rep. ID/WG 151/4 U.N. Indus. Dev. Org., Vienna, Austria
Pratt, G. H.
    1968. An appraisal of the drying of timber with the aid of dehumidifiers. Rep. Seminar on Dehumidifiers for Timber Drying, For. Prod. Res. Lab., Princes Risborough, Aylesbury, Bucks., England.
Read, W. R., A. Choda, and P. I. Cooper
    1974. A solar timber kiln. Solar Energy 15(4): 309–316.
Souter, G. R.
    1971. Timber drying by dehumidification. Timber Trade J. Annual Spec. Issue, Sawmilling and Woodwork. Sect. S/23–28.
Ullevalseter, R. O.
    1971. Lumber drying by condensation with the use of refrigerated dew point. Inst. Wood Technol., Vollebekk (Norway). p. 64.
Wengert, E. M.
    1971. Improvements in solar dry kiln design. USDA For. Serv. Res. Note FPL–0212. For. Prod. Lab., Madison, Wis.

Glossary

Air drying. The process of drying green lumber or other wood products by exposure to prevailing natural atmospheric conditions outdoors or in an unheated shed.

Air drying calendar. A table showing the number of effective air drying days each month of the year in a specific area.

Air drying efficiency. Operation of the air

drying process to reduce green wood to the air-dried moisture content level with least cost in energy, time, and money.

Good air drying month. Thirty consecutive days in which daily mean temperatures are over 45 degrees F.

Board. Yard lumber that is less than 2 inches thick and 2 or more inches wide; a term usually applied to 1-inch thick stock of all widths and lengths.

Check. A separation of the wood fibers within or on a log, timber, lumber, or other wood product resulting from tension stresses set up during drying, usually the early stages of drying. Surface checks occur on flat faces of boards; end checks on ends of logs, boards, or dimension parts.

Chemical seasoning. The application of a hygroscopic chemical (e.g. sodium chloride) to green wood for the purpose of reducing defects, mainly surface checks, during drying. The chemical may be applied by soaking, dipping, spraying with aqueous solutions, or by spreading with the dry chemical and bulk piling.

Collapse. Flattening or buckling of the wood cells during drying resulting in excessive or uneven shrinkage plus a corrugated surface.

Dry kiln. A room, chamber, or tunnel in which the temperature and relative humidity of air circulated through parcels of lumber, veneer and other wood products can be controlled to govern drying conditions.

Drying. The process of removing moisture from wood to improve its serviceability in use.

Drying rates. The loss of moisture from lumber or other wood products per unit of time. Generally expressed in percentage of moisture content lost per hour or per day.

Edge piling. In air drying, stacking of wood products on edge. so that the broad face of the item is vertical. In kiln drying, stacking of lumber on edge for drying in kilns with vertical air circulation.

End coating. A coating of moisture-resistant material applied to the end-grain surfaces of green wood such as logs, timbers, boards, squares, etc. to retard end drying and consequent checking and splitting or to prevent moisture loss from the ends of the drying samples.

End pile. In air drying, stacking of green lumber on end, and inclined, in a long fairly narrow row, the layers separated by stickers.

Equilibrium moisture content. The moisture content at which wood neither gains or looses moisture when surrounded by air at a given relative humidity and temperature. EMC is frequently used to indicate potential of an atmosphere to bring wood to a specific MC during a drying operation.

Flue. In stacking lumber or other wood products for air drying, a vertical space 6 inches or less in width in the center of the pile and extending the length of the pile, intended to facilitate circulation of air within the pile.

Hardwoods. Generally one of the botanical groups of trees that have broad leaves in contrast to the conifers or softwoods. The term has no reference to the actual hardness of the wood.

Heartwood. The inner layer of a woody stem wholly composed of nonliving cells and usually differentiated from the outer enveloping layer (sapwood) by its darker color. It is usually more decay resistant than sapwood and more difficult to dry.

Honeycombing. In lumber and other wood products, separation of the fibers in the interior of the piece, usually along the wood rays. The failures often are not visible on the surfaces, although they can be the extensions of surface and end checks.

Humidity. The moisture content of air.

Relative humidity. Under ordinary temperatures and pressures, it is the ratio of the weight of water vapor in a given unit of air compared to the weight that the same unit of air is capable of containing when fully saturated at the same temperature. More generally, it is the ratio of the vapor pressure of water in a given space compared with the vapor pressure at saturation for the same dry-bulb temperature.

Kiln. A chamber or tunnel used for drying and conditioning lumber, veneer, and other wood products in which the temperature and relative humidity of the circulated air can be varied.

Lumber. The product of the sawmill and planing mill not further manufactured than by sawing, resawing, passing lengthwise through a standard planing machine, cross cutting to length, and matching.

Boards. Yard lumber less than 2 inches thick and 1 or more inches wide.

Common lumber. A classification of medium and low-grade hardwood lumber and/or softwood lumber suitable for general construction but not suitable for finish grade.

Finish lumber. A collective term for upper grades of lumber suitable for natural or stained finishes.

Moisture content. The amount of water contained in the wood, usually expressed as a percentage of the weight of the oven-dry wood.

Moisture content classes.

Air dried. Wood having an average moisture content of 25 percent or lower, with no material over 30 percent.

Green. Freshly sawed wood or wood that essentially has received no formal drying.

Kiln dried. Dried in a kiln or by some other refined method to an average MC specified or understood to be suitable for a certain use, such an average generally being 10 percent or below for hardwoods.

Permeability. The ease with which a fluid flows through the porous material (wood) in response to pressure.

Pile. In air drying, stacking lumber layer by layer, separated by stickers, on a supporting foundation. Also, stickered unit packages by lift truck or crane, one above the other on a foundation and separated by bolsters.

Predrying. A wood drying process carried out in special equipment before kiln drying.

Predrying treatments. Special measures before or early in drying to accelerate drying rate, to modify color, or to prevent checks and other drying defects.

Chemical. Impregnating the outer zone of green lumber with hygroscopic and sometimes bulking chemicals to about one-tenth the thickness to reduce surface checking during drying. Involves control of relative humidity at a value below the RH that is in equilibrium with the saturated chemical solution.

Polyethylene glycol. Deeply impregnating green lumber with polyethylene glycol 1000 (PEG) to retain the green dimension during drying and indefinitely thereafter, minimizing shrinkage, swelling, and warp.

Precompression. Momentarily compressing green lumber transversely about 7.5% to permit drying by a severe schedule and improve drying behavior.

Prefreezing. The freezing and subsequent thawing of green wood before drying to increase drying rate and decrease shrinkage and seasoning defects.

Steaming. Subjecting green wood to saturated steam at or close to 212° F to accelerate drying, achieve a desirable color, or both.

Surface treatment. Applying a salt or sodium alginate paste to the surface of green wood to help prevent checking as the wood is dried by other means.

Sapwood. In wood anatomy, the outer layers of the stem that in the living tree contain living cells and reserve materials, e.g., starch. The sapwood is generally lighter in color than the heartwood.

Shrinkage. The contraction of wood fibers caused by drying below the fiber saturation point, Shrinkage—radial, tangential and volumetric—is usually expressed as a percentage of the dimension of the wood when green.

Longitudinal shrinkage. Shrinkage of wood along the grain.

Radial shrinkage. Shrinkage across the grain, in a radial-transverse direction.

Tangential shrinkage. Shrinkage across the grain, in a tangential-transverse direction.

Volumetric shrinkage. Shrinkage of wood in volume.

Spacing. Row spacing. Spaces between rows of piles in the yard.

Sticker spacing. Distance between adjacent stickers in a pile, kiln truckload, or a stickered unit package of lumber.

Sticker. A wooden strip, or its substitute, placed between courses of lumber or other wood products, in a pile, unit packages, or kiln truckload, at right angles to the long axis of the stock, to permit air to circulate between the layers.

Storage. Bulk or stickered piling of air or kiln-dried wood products with protection from the weather in accordance with the desired level of moisture content,; protection may be tarpaulins, or open, closed, or closed and heated sheds.

Tangential surface. Surface tangent to the growth rings; a tangential section is a longitudinal section through a log perpendicular to a radius. Flat-grained lumber is sawed tangentially.

Tier. In air drying or kiln drying, a stack of packages of lumber or other wood products in vertical alinement. Also refers to sticker alinement from layer to layer.

Transverse. The directions in wood at right angles to the wood fibers or across the grain. A transverse section is a section through a tree or timber at right angles to the pith.

Unit package. Lumber or other wood products that have been assembled into a parcel for handling by a crane, carrier, or forklift truck.

Vapor barrier. In kiln drying, a material with a high resistance to vapor movement that is applied to dry kiln surfaces to prevent moisture migration.

Vent. In kiln drying, an opening in the kiln roof or wall that can be opened and closed to control the wet-bulb temperature within the kiln.[23]

NOTES

1. Adapted from Nancy K. Mello, Jack H. Mendelson, Mark P. Bree, and Scott E. Lukas, "Buprenorphine Suppresses Cocaine Self-Administration by Rhesus Monkeys," *Science*, 245 (1989), pp. 859–862.

2. Adapted from Constance Sorrentino, "The Changing Family in International Perspective," *Monthly Labor Review*, 113:3 (1990), pp. 42–42.

3. Adapted from Jonathan A. Showstack, Mary Hughes Stone, and Steven A. Schroeder, "The Role of Changing Clinical Practices in the Rising Costs of Hospital Care," *The New England Journal of Medicine*, 313:19 (1985), p. 1201.

4. Adapted from *The Safe Food Book: Your Kitchen Guide*, U.S. Department of Agriculture, July 1984, p. 8.

5. Adapted from "Dietary Fads and Frauds," *The Surgeon General's Report on Nutrition and Health 1988*, U.S. Department of Health and Human Services, Publication No. 88-50210, pp. 702–703.

6. Adapted from Aline McKenzie, "When Chunks Collide," *Science Notes*, 15:2 (1989), p. 3.

7. Adapted from Richard W. Chapman, Salvatore Tozzi, and William Kreutner, "Antibronchoconstrictor Activity of the Intracellular Calcium Antagonist HA 1004 in Guinea Pigs," *Pharmacology*, 37 (1988), p. 190.

8. Adapted from *Simple Home Repairs Outside*, U.S. Department of Agriculture, 1986, Program Aid Number 1193, p. 12.

9. Ibid., p. 13.

10. Sorrentino, p. 44.

11. Ibid., pp. 42–43.

12. Ibid., pp. 42–44.

13. Richard P. Feynman, *QED: The Strange Theory of Light and Matter* (Princeton, N.J.: Princeton University Press, 1985), p. 13.

14. Rudolph W. Koster and Arend J. Dunning, "Intramuscular Lidocaine for Prevention of Lethal Arrhythmias in the Prehospitalization Phase of Acute Myocardial Infarction," *The New England Journal of Medicine*, 313:18 (1985), p. 1106.

15. Adapted from *1984 Yearbook of Agriculture Animal Health*, U.S. Department of Agriculture, pp. 404–410.

16. *The Safe Food Book*, pp. 8–17.

17. Ibid., pp. 17–23.

18. Adapted from Roger M. Waller, *Ground Water and the Rural Homeowner*, U.S. Department of the Interior, undated, pp. 8–12.

19. *Simple Home Repairs Outside*, p. 14.

20. Adapted from *Simple Home Repairs Inside*, U.S. Department of Agriculture, 1986, Program Aid Number 1034, pp. 15–16.

21. Adapted from *Radon Reduction Methods: A Homeowner's Guide*, U.S. Environmental Protection Agency, September 1987.

22. Sorrentino, pp. 48–53.

23. Adapted from John M. McMillen and Eugene M. Wengert, *Drying Eastern Hardwood Lumber*, U.S. Department of Agriculture, Agricultural Handbook No. 528, pp. 61–67.

INDEX

4290828

Made in the USA
Lexington, KY
11 January 2010